www.KnowledgePublications.com

SAVE HARMLESS AGREEMENT

Because use of the information, instructions and materials discussed and shown in this book, document, electronic publication or other form of media is beyond our control, the purchaser or user agrees, without reservation to save Knowledge Publications Corporation, its agents, distributors, resellers, consultants, promoters, advisers or employees harmless from any and all claims, demands, actions, debts, liabilities, judgments, costs and attorney's fees arising out of, claimed on account of, or in any manner predicated upon loss of or damage to property, and injuries to or the death of any and all persons whatsoever, occurring in connection with or in any way incidental to or arising out of the purchase, sale, viewing, transfer of information and/or use of any and all property or information or in any manner caused or contributed to by the purchaser or the user or the viewer, their agents, servants, pets or employees, while in, upon, or about the sale or viewing or transfer of knowledge or use site on which the property is sold, offered for sale, or where the property or materials are used or viewed, discussed, communicated, or while going to or departing from such areas.

Laboratory work, scientific experiment, working with hydrogen, high temperatures, combustion gases as well as general chemistry with acids, bases and reactions and/or pressure vessels can be EXTREMELY DANGEROUS to use and possess or to be in the general vicinity of. To experiment with such methods and materials should be done ONLY by qualified and knowledgeable persons well versed and equipped to fabricate, handle, use and or store such materials. Inexperienced persons should first enlist the help of an experienced chemist, scientist or engineer before any activity thereof with such chemicals, methods and knowledge discussed in this media and other material distributed by KnowledgePublications Corporation or its agents. Be sure you know the laws, regulations and codes, local, county, state and federal, regarding the experimentation, construction and or use and storage of any equipment and or chemicals BEFORE you start. Safety must be practiced at all times. Users accept full responsibility and any and all liabilities associated in any way with the purchase and or use and viewing and communications of knowledge, information, methods and materials in this media.

www.Knowledgeublications.com

presents

The Free Alcohol Fuel Compendium

Volume 2

including

Two-Stage, Dilute Sulfuric Acid Hydrolysis of Wood and Wood Hydrolysis for Sugar Production

---

ISBN: 978-1-60322-068-2

United States Department of Agriculture

Forest Service

Forest Products Laboratory

General Technical Report FPL-45

Two-Stage, Dilute Sulfuric Acid Hydrolysis of Wood:

An Investigation of Fundamentals

John F. Harris
Andrew J. Baker
Anthony H. Conner
Thomas W. Jeffries
James L. Minor
Roger C. Pettersen
Ralph W. Scott
Edward L. Springer
Theodore H. Wegner
John I. Zerbe

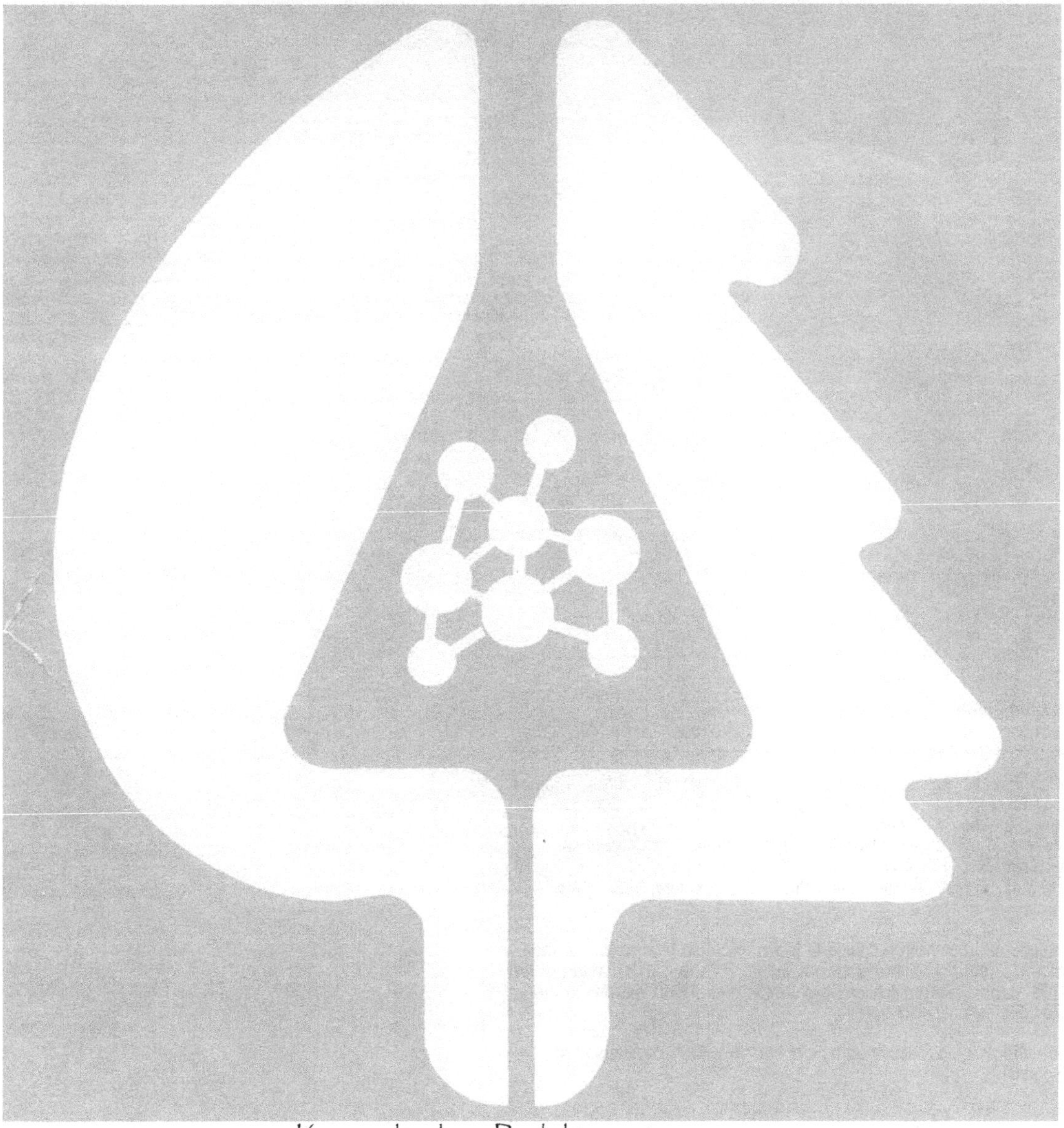

Abstract

This paper presents a fundamental analysis of the processing steps in the production of methanol from southern red oak (*Quercus falcata* Michx.) by two-stage dilute sulfuric acid hydrolysis. Data for hemicellulose and cellulose hydrolysis are correlated using models. This information is used to develop and evaluate a process design.

Keywords: Southern red oak hemicellulose, cellulose acid hydrolysis, reaction kinetics, yields, process, xylose, glucose, ethanol, furfural

Conversion of Units

Multiply	By	To obtain
1 kilogram (kg)	2.205	pounds
1 tonne = 1,000 kg	1.102	tons (U.S.)
1 joule (J)	0.2388	calories
1 kilojule (kJ)	0.9478	Btu
1 kilopascal (kPa)	0.1450	psi
1 kilojoule/kilogram (kJ/kg)	0.4299	Btu/pound
1 millimeter	0.03937	inches

October 1985

Harris, John F.; Baker, Andrew J.; Conner, Anthony H.; Jeffries, Thomas W.; Minor, James L.; Pettersen, Roger C.; Scott, Ralph W.; Springer Edward L; Wegner, Theodore H.; Zerbe, John I. Two-stage, dilute sulfuric acid hydrolysis of wood: An investigation of fundamentals. Gen. Tech. Rep. FPL-45. Madison, WI: U.S. Department of Agriculture, Forest Service, Forest Products Laboratory; 1985. 73 p.

A limited number of free copies of this publication is available to the public from the Forest Products Laboratory, P.O. Box 5130, Madison, WI 53705. Laboratory publications are sent to over 1,000 libraries in the United States and elsewhere.

The Laboratory is maintained in cooperation with the University of Wisconsin.

Table of contents

Two-Stage, Dilute Sulfuric Acid Hydrolysis of Wood: An Investigation of Fundamentals[1]

John F. Harris, Chemical Engineer
Andrew J. Baker, Chemical Engineer
Anthony H. Conner, Chemist
Thomas W. Jeffries, Microbiologist
James L. Minor, Chemist
Roger C. Pettersen, Chemist
Ralph W. Scott, Chemist
Edward L. Springer, Chemical Engineer
Theodore H. Wegner, Chemical Engineer
John I. Zerbe, Technologist

Forest Products Laboratory, Madison, Wis.

Introduction

Of the various methods by which wood can be converted to energy, the simplest and mast efficient is direct combustion. However, because of the continuing need to reduce liquid petroleum imports, there has evolved a strong national cummitment to produce energy from wood in the farm of ethanol. In keeping with its historical role in researching the production of chemicals and energy from wood, the Forest Products Laboratory has undertaken selecting, describing, and improving the most promising process from among those that can conceivably be commercialized in the near future. We believe that process is two-stage dilute-acid hydrolysis.

The process, in its simplest outline, is shown in figure 1. Wood chips, impregnated with a dilute sulfuric acid solution and drained of all interstitial liquid, are charged to the first stage. Here they are heated with direct steam, resulting in the hydrolysis of most of the hemicelluloses, and then discharged to washers. After being washed free of the material solubilized in the first stage, the lignocellulose is reimpregnated with acid and charged to the second stage. As with the first stage, the liquid content of the second-stage charge is kept to a minimum. Conditions here are sufficient to hydrolyze the resistant cellulose. The resulting mixture is discharged to washers where the glucose solution is separated from the lignin residue.

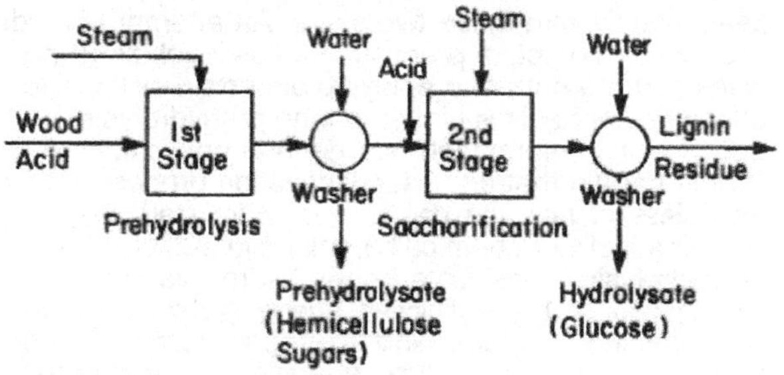

Figure 1.—Simple schematic of the two-stage hydrolysis process. (ML84 5484)

[1]This report was written as a result of research undertaken by various projects at the Forest Products Laboratory (FPL) through a cooperative agreement with the Tennessee Valley Authority, Muscle Shoals, Ala. The research team was led by John F. Harris and received administrative advice and guidance from John I. Zerbe, Program Manager, and Andrew J. Baker, Chemical Engineer, of the Energy Research, Development, and Application Program at the FPL.

This process was selected for investigation because it was thought that: (1) The process could be carried rapidly to commercial operation, on the basis of the amount of information available on the use of dilute sulfuric acid as a hydrolysis medium. (2) The separate stages for hydrolyzing the hemicelluloses and cellulose would result in high yields and high-purity products. Because much of the liquid is removed before each of the hydrolysis steps, (3) the energy consumption would be minimized, and (4) the resulting sugar solutions would be more concentrated.

Some developmental work on a similar process was carried out in Sweden during World War II and a short description of the results presented at a United Nations conference in 1952 (Cederquist 1954). Although the results were encouraging, the process was clearly considered to have been a wartime expedient, and work was discontinued when the war ended. No attempt had been made to deal with effluents, to complete the design, or to optimize the process. The wood used was spruce (*Picea excelsa* Link) rather than the hardwoods-southern red oak (*Quercus falcata* Michx.), in particular-that are of greater importance in the American economy. The approach taken in this research was shaped by the large amount of fundamental data available from previous research at the Forest Products Laboratory. Existing information was believed sufficient to permit a fairly accurate design of the, process-that is, the more important components could be modeled and brought info a harmonious whole. We also did experimental work to validate or modify the assumptions and data used in modeling. Our purpose was to recommend processing conditions pertaining to each element, and to estimate yields, energy requirements, and other pertinent process information.

The major process elements are the first-stage prehydrolysis and second-stage hydrolysis. Since these are within the area of our expertise, the report deals mainly with these two steps. An attempt is made to outline a complete process, but this involves using less accurate data and applying uncertain skills. The principal emphasis is placed on the prehydrolysis, because this information has general application. It is almost certain that any saccharification process, regardless of how the cellulose is hydrolyzed (enzymatically or chemically), will have a dilute acid prehydrolysis stage. Data on the hydrolysis and degradation of hemicelluloses during prehydrolysis are, also of value in establishing the coproduction of pulp and chemicals such as furfural and acetic acid or the production of animal feedstuffs including molasses and proteins. Prehydrolysis studies were extended to the point of simulating industrial conditions in a small digester, whereas experimental work on the second-stage hydrolysis was restricted to laboratory scale. The second-stage studies were delayed since they required residue from the first stage. Since no representative hydrolysis solutions were available from the lignocellulose hydrolysis, no fermentation studies with it were done, but the known contaminants of the hydrolysate are being tested for toxicity in yeast fermentation.

The project has spawned several related fundamental studies not essential to the current process design. They include investigations of xylose metabolism, cellulose hydrolysis, sugar-degradation and deacetylation kinetics. We plan to continue these studies and release the results through technical journals.

This report presents information currently available to help carry the process into pilot-plant development. It also contains facts needed to evaluate the process at its present state of development and to select areas for study that would lead to improvement. Considerable attention is given to a discussion and analysis of the underlying physical and chemical mechanisms. This is important in obtaining an appreciation of the many problems and the possibilities for future development.

First Stage (Prehydrolysis)

Hemicellulose Removal

Xylan removal.—The removal of xylan from a hardwood prehydrolyzed under conditions of constant temperature, acidity, and liquid-to-solid ratio (L/S) can best be described by a typical curve on a semilogarithmic plot (fig. 2). The first portion of the curve, to about the point of 60% removal, is quite linear; the slope is essentially constant. Thereafter, the removal rate continuously decreases until it reaches a minimum value equal to the hydrolysis rate of cellulose at the particullar contitions employed. This is deduced from the fact that, on prolonged prehydrolysis, the residue reaches a constant value (Harris et al. 1963) indicating that a small portion (1-2%) of the xylan is intimately associated with the resistant cellulose.

Throughout the discussion that follows, the term "rate constant" will be used to refer to the absolute value of the slope of this xylan weight removal curve. It should be understood that this rate constant is dependent on, but not directly related to, the rate constants for individual bond cleavage. The rate of dissolution involves the solubilities of various oligomers as well as the rate of bond cleavage. Also the the chemical structure of the polymeric solid must be changing throughout the course of hydrolysis since the proportion of resistant bonds must be increasing. It is possible that the decreasing removal rate is diffusion not chemically controlled.

Springer (1966) investigated the effect of acid concentration (HCl) and temperature on the initial rate constant for the removal of xylan from aspen wood and found that it could be described by the equation:

$$\log_{10}(k/CH) = 15.083 - 6171.3/T + 0.22219(CH) \quad (1)$$

where

T = absolute temperature (°K)
CH = molarity of hydrogen ion, $[H^+]$
k = rate constant (min^{-1})

Springer and Zoch (1968) also compared the initial rate constants for several wood species and found only small variation from species to species. The minor variations corresponded with differences in the proportion of 4-O-methylglucuronic acid side chains in the xylans.

Reasons for the reduction In removal rate after 50-60% of the xylan is removed are unknown. Neither the position nor associated slopes of points on the later portion of the curve In figure 2 can be predicted. Consequently, it is necessary to collect experimental data in the region of interest. This region for a prehydrolysis aimed at maximum xylose recovery Is between 90 and 95% xylan removal.

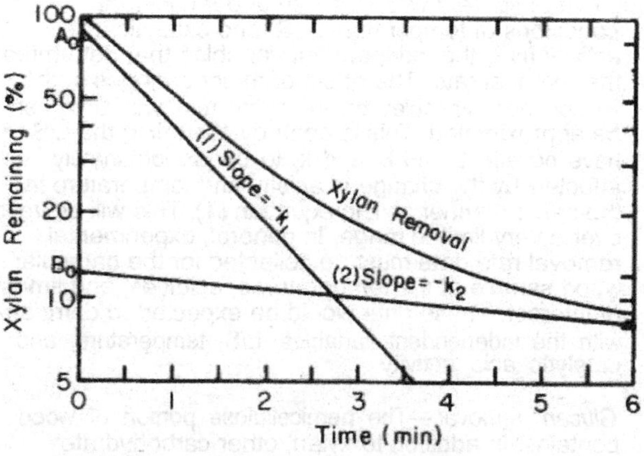

Figure 2.—Xylan removal during prehydrolysis. Curves (1) and (2) are linear components of the xylan removal curve. (ML84 5478)

It is convenient to have an analytical expression for the xylan removal curve. The derivation of the functional form that was used In this study is described below and illustrated (fig. 2). The removal curve is assumed to be the sum of two functions, each linear on a semilogarithmic plot:

XR = xylan remaining in residue, % of original

$$= A_o\exp(-k_1t) + B_o\exp(-k_2t) \quad (2)$$

where t = time.

Since the plot is presented as percentage removal, at t = 0

$$XR = A_o + B_o = 100 \quad (3)$$

so that:

$$XR = A_o\exp(-k_1t) + (100 - A_o)\exp(-k_2t) \quad (4)$$

An experimental data set can be used with equation (4) to obtain the values of A_0, k_1, and k_2 for the best fit. These-values substituted into equation (2) yield the desired analytical expression. One should not be tempted to attach any fundamental significance to this correlation. It is readily apparent from figure 2 that the values of the parameters will depend on the particular range In which the data were collected.

It has been assumed in the above discussion that the curve of figure 2 was obtained under constant conditions of temperature, L/S, and catalytic acid activity [H^+], the independent variables that determine the removal rate. The effect of minor changes in the independent variables on the xylan removal curve can be approximated. This is done by assuming the L/S to have no effect, and k_1 and k_2 to be proportionately affected by the change in acidity and temperature in the same manner as k in equation (1). This will be valid over a very limited range. In general, experimental removal rate data must be collected for the particular wood sample at the temperatures, acidities, and times of interest. These data would be expected to correlate with the independent variables, L/S, temperature, and catalytic acid activity.

Glucan removal.—The hemicellulose portion of wood contains, in addition to xylan, other carbohydrate materials such as galactoglucomannan and arabinogalactan. These are hydrolyzed and removed during prehydrolysis in a manner similar to the removal of xylan. The resulting sugars–glucose, mannose, arabinose, and galactose–form a sizable portion of the total carbohydrate content of the prehydrolysate even in the case of southern red oak where xylose is predominant. Rather more glucose appears in the hydrolysate than could have originated from the galactoglucomannan. Although there is probably some amorphous glucan associated with the hemicelluloses (Rydholm 1965), most of the additional glucose must originate from the cellulose fraction. Typically, cellulose hydrolysis indicates than cellulose has an easily hydrolyzable component that is removed at a rate comparable to that of hemicellulose (Millett et al. 1954). It is this easily hydrolyzable component of the cellulose that was solubilized during prehydrolysis.

Prehydrolysis data. — Some data[2] on the composition of prehydrolysis residues were available when this project was begun. Baker and Krcmar had measured the lignin, glucan, xylan, and mannan contents of residues from several species, including southern red oak, using conditions of possible commercial interest. Xylan removal curves were obtained for 170° and 190° C at several acid levels—0.1, 0.4, and 1.6%. Time intervals were adjusted so that the extent of xylan removal was always more than 80% for any set of conditions.

These data were used, as described later, to estimate yields and thus set the experimental range of the variables to be used in further studies.

Catalyst concentration. — Determining the catalytic activity presents a problem. Hydrogen ion activity is usually a questionable quantity in any kinetic study because it is not exactly equal to the concentration of H^+, it is an unknown function of temperature, and it varies somewhat, again in some unknown way, with the concentration of other components in solution. In hydrolyzing wood there is an additional factor–the neutralizing capacity of the substrate. At low pH the wood neutralizes some of the added acid. Thus, the L/S, concentration of the added acid, and the neutralizing capacity all influence the catalyst concentration. The neutralizing capacity will have a minor effect of high L/S or high added acid concentrations. At low L/S and low acid concentrations — conditions of interest to the process presented here — it is a major factor to consider.

All woods are slightly acidic; depending on the species, the pH of green wood is between 4 and 6. When wood is heated, acetyl groups are released as acetic acid, decreasing the pH to below that of the green wood. The presence of these weakly acidic organic does not enter into what we have referred to above as the neutralizing capacity of the wood because they are not ionized at low pH. The principal components that are effective in decreasing the acidity are salts of 4-O-methylglucuronic acid. Since the pK (pK = - log K; K = dissociation constant) of this acid is about 3, at the pH levels of hydydrolysis (~1.5) the organic salt is completely converted to the undissociated acid. Thus each equivalent of salt neutralizes an equivalent of acid. Not all of the 4-O-methylglucuronate groups are present in the wood as salts; some are esterified. It can be deduced from the measurements on the southern red oak used in this study (table 1) that about 65% are in salt form. The inorganics associated with the 4-O-methylglucuronate salts are the major cationic components found in the wood ash, and, consequently, the ash content is an indicator of the neutralizing capacity. There are other inorganics in the ash that do not contribute to the neutralizing capacity of the wood; these are primarily silica and inorganic salts, which are present in minor quantities.

---

[2]Baker, A. J.; Krcmar, G. F. Kinetics of the prehydrolysis of wood. Madison, WI: U.S. Forest Service, Forest Products Laboratory; January 1956. 17 p. Unpublished report.

Table 1.—Analysis of southern red oak (Quercus falcate Michx.) wood samples

Component	Sample number			
	1[1]	2	3[1]	4
	%			
Glucan	40.3	38.8	37.8	41.8
Mannan	2.9	2.7	2.1	2.2
Xylan	19.3	19.1	18.4	18.7
Galactan	([2])	1.4	1.1	.8
Arabanan	([2])	1.4	.71	.82
Uronic anhydride	2.9	2.7	3.3	2.7
Acetyl	3.7	3.9	4.3	3.5
Lignin	21.8	22.2	21.9	19.5
Ash	.24	.50	.72	.29
Extractives	5.5	6.0	6.7	6.6
Total	96.6	98.7	97.0	96.9

[1] Sample 1 was used for ampoule studies; sample 3 for digester studies.

[2] The analytical procedure used does not distinguish galactan from glucan nor arabanan from mannan.

A direct approach to measuring neutralizing capacity was unsuccessful. It was attempted by adding a known quantity of H_2SO_4 to a weighed wood sample, removing an aliquot, and back-titrating with NaOH. The presence of colored extractives prevented the use of an indicator, and the weak organic acids present resulted in an uninterpretable titration curve. Two alternatives were tried.

In one procedure the wood was ashed, an excess of strong acid added, and then back-titrated to the phenolphthalein end point. The equivalents of base in the ash should be the same as that in the wood. Both inorganic anions and cations should be carried quantitatively through ashing. Salts of strong acids are neutral, and neither their presence nor loss due to volatility affect neutralization capacity.

A second alternative was an analysis far the elements; from these data, the neutralization capacity was calculated. Elemental analysis can be done directly an the wood with neutron activation or by solubilizing the wood or the wood ash and assaying the solution, preferably using atomic emission and absorption spectrometry. Neutron activation was not used because of the lack of a suitable calcium standard. Data from the titration of the ash and from atomic spectrometry agreed well. The atomic spectroscopy was done in two laboratories using different procedures for solubilizing inorganic ions (table 2). It was concluded that the simple procedure of ashing and titrating is satisfactory. In the absence of silica, an approximation of the neutralizing capacity can be made by assuming the ash to consist entirely of $CaCO_3$ and K_2CO_3 in the weight ratio 3:2. There is considerable variation in the ash content of different wood samples of the same species (table 1), Sapwood was found to have a much higher ash content than heartwood. For one particular log, these values were, respectively, 1.38 and 0.99%; in another, 0.86 and 0.43%.

The procedure used throughout this work to calculate the concentration of hydrogen fan or the desired acid concentration, at various conditions, is given in Appendix A.

Studies were done to test the assumed effect of the ash constituents an the removal rate of xylan from southern red oak (table 1, sample 1). These and other bench-scale studies were done using wood samples in 5-mm glass ampoules; the experimental technique has been described elsewhere (Springer 1961; Springer

Table 2.—Comparative neutralization capacity of southern red oak,[1] by three methods

Method	Elemental analysis (atomic emission and absorption spectrometry)									Neutralizing capacity
	Ca	K	Mg	Na	Mn	Fe	Cu	S	P	
	ppm wood									meq/kg wood
Elution with HC1[2]	1,340	1,410	167	32	55	109	14	—	—	114
Digested with nitric-perchloric[3]	1,430	1,380	160	104	52	162	9.4	120	80	121
Ashing (TAPPI 1982)[4] and titrating	—	—	—	—	—	—	—	—	—	116

[1]Sample 3 from table 1, ground to 40-80 mesh.

[2]Cations removed from wood by column elution with HC1.

[3]Wood sample digested with nitric-perchloric acid mixture.

[4]Ash content = 0.72% by muffle furnace analysis. Ash neutralizing capacity = 16.17 eq/kg of ash by phenolphthalein analysis.

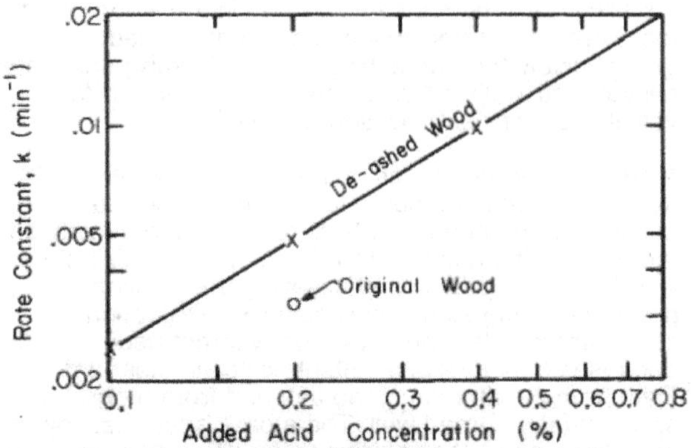

Figure 3.—A comparison of initial xylan removal rate constants for original and de-ashed wood. (ML84 5479)

et al. 1963). A rather law temperature of 120° C was. chosen to obtain the best experimental accuracy. The L/S was 3. As previously discussed, the removal rate for the first 60% of the xylan is linear on a semilogarithmic plot, and the initial weight loss can be represented by a kinetic constant, k, similar to a first-order reaction rate constant. This constant is nearly proportional to acid concentration when the solution is dilute (eq, (I)). Samples of oak wood were ground to pass 40 mesh, and the ash constituents were nearly completely removed by washing with dilute hydrochloric acid. The initial rate constants far this "de-ashed" wood were determined at various sulfuric acid concentrations (the line in fig. 3). The rate constant for the original wood (40 mesh) was measured in 0.2% H_2SO_4 using the same conditions (120° C, L/S = 3) (the single point in fig. 3).

The horizontal displacement of the point from the line for "de-ashed" wood represents the lass in acidity resulting from the neutralizing capacity of the wood. The effective neutralizing power calculated from the difference in acid strength is 31 milliequivalents per kg (meq/kg) of wood. From the elemental, ash. composition, it was calculated that the wood should have a neutralizing capacity of 37 meq/kg; this is in reasonable agreement with the value measured kinetically. This implies that a good approximation to catalytic activity can be made from the ash content and its composition. It should be remembered that this was only experimentally verified over the 60% xylan removal range. in this study, which used finely ground wood in small ampoules, nearly all the ash constituents were effectively neutralized; in subsequent work with chips, to be discussed later, it was found that this was not the case. Studies determining the effect of ash constituents an prehydrolysis rate are being continued.

Removal of other components. — Acetyl and uronic anhydride substituents of the hemicellulase xylan were also removed during prehydrolysis. These were bath sizable components of the wood and came into process considerations. Often these components are not included in the wood analysis. Approximate values can be estimated by assuming that the xylose anhydride units in the wood are 60% substituted with acetyl groups and 13% with uranic anhydride groups (Rydholm 1965). There is little quantitative information regarding their rates of removal, the manner in which they are removed, or the subsequent reactions they undergo during hydrolysis.

Acetyl is not as readily removed from wood as is xylose anhydride. When only 10% of the xylan remains, as much as 25%. of the original acetyl is still in the residue. This finding suggests that the easily removable xylan is not as highly substituted with acetyl groups as the more resistant xylan. Supporting this conclusion, studies have shown that most of the acetyl groups are released from solubilized xylan fragments and not from *in situ* xylan.

It has been surmised that xylan is bound to lignin through the uronic acid side group (Lundquist et al. 1980). it could be assumed that such banding might tend to deter the removal of uranic groups. Also, from the structure of the xylan polymer, one would predict, an the basis of glycoside hydrolysis rates (Harris 1975), that the removal of the uranic groups would be somewhat slower than removal of the anhydraxylose units. Our data indicate that the effect of these two factors is small since the uronic units are only slightly more resistant to removal than the xylose units. Most of the uranic depletion is accounted for as the acid in the hydrolysate, indicating that it was removed by hydrolysis and did not degrade *in situ.*

Most of the extractives and a small portion of the lignin are solubilized during prehydrolysis. The lignin undergoes extensive change, but not much is solubilized. Same lignin bonds are readily hydrolyzed and during the initial period of reaction significant depolymerization occurs. However, the depolymerized material is quite reactive and recombines to form high molecular weight, insoluble material with properties quite unlike those of the original lignin. Same demethoxylation occurs releasing methanol, which is a contaminant that must be considered in process design.

Table 3.—Calculated maximum xylose yields from hydrolysis of southern red oak[1]

Temperature	Acid		Reaction time	Xylan remaining	Maximum xylose yield[3]	Loss
	Added	Effective[2]				
°C	%		min	%		
170	0.1	0.064	19.8	13.0	81.3	5.7
	.4	.36	4.6	11.0	83.2	5.8
	1.6	1.56	1.4	6.5	86.3	7.2
190	.1	.064	3.5	10.3	85.1	4.6
	.4	.36	2.6	10.2	77.1	12.7

[1]Baker, A. J.; Krcmar, G.F. Reference from text footnote 2, page 4.

[2]Calculated from, an assumed wood ash content of 0.3% and L/S = 10.

[3]Calculated, % of potential.

Xylose and Furfural Yields

Data on the decomposition rate of xylose, covering a broad range of. temperature, acidity (H_2SO_4), and xylose concentration, are available (Root 1956; Root et al. 1959). The first-order rate constant is correlated by the expression:

$$k = 2.72 \cdot \alpha \cdot \delta \cdot \gamma \cdot CA \cdot \exp[-35.7(473.1-T)/T] \qquad (5)$$

where

α = $\alpha(CX)$, a function of xyloee molarity, where CX = xylose molarity
δ = $\delta(T)$, a function of temperature
γ = $\gamma(T,CA)$, a function of temperature and acid normality
CA = acid normality
T = absolute temperature (°K)

No functional forms are available for a, d, g, but their relationships to the independent variables are given in tabular form (Root 1956; Root et al. 1959).

This information, along with the xylan removal rate data, can be used to calculate the xylose content of the solution in contact with the wood as a function of the time of prehydrolysis. Numerical integration is required because there is no analytical expression for k, which varies as the reaction proceeds. Since xylose is being liberated, the value of CX varies, changing the values of a. The change in xylose concentration also indirectly affects the acid normality since the volume of the solution changes. Thus, g also varies throughout the reaction.

Calculation of xylose yields in the manner above involves many assumptions. The most unreliable of these is probably the assumption that the xylan is released into solution entirely as monomeric xylose, disregarding the presence of oligomers. Since decomposition of the sugar proceeds from the reducing end group, the protection afforded by combination in oligomers would decrease the xylose loss. Experimentally, we have found that, In dilute acid solutions at maximum yield, less than 30% (often 10-20%) of the total xylose content of the solution is present in oligomeric form.

The residue data on southern red oak available from previous work[2] were used to calculate xylose yields (table 3). Calculated maximum yields are all above 80% except for the condition 190° C and 0.4% H_2SO_4. Reexamination of this set of the data and of the experimental procedure led to the conclusion that this point was unreliable. From the remainder, it was concluded that yields in excess of 80% would be obtained, and that increasing temperature and acidity both tend to increase maximum yields. In these experiments, data were collected an the residue composition only; no analysis was made on the accompanying hydrolysates. Thus it is not possible to compare the calculated values with experimental data.

To remedy this, data on both residues and hydrolysates were obtained for a variety of hydrolysis conditions with southern red oak (table 1, sample 1) using ampoules (see Appendix G) (table 4). Calculations of xylose yields can be made using these data and the computed maximum yields compared to those measured (table 5). The calculated xylose yields are all in excess of 80%; they indicate a modest increase with increasing temperature but a slight decrease with increasing acid. The measured values are, with one exception, somewhat greater than the calculated values; this may result from the protection afforded by the presence of oligomers noted earlier. The experimental yields indicate that the maximum yields increase with increasing acid strength and temperature.

Table 4.–Prehydrolysis fractionation of southern red oak (from table 1, sample 1), L/S = 4

Temperature	Added acid concentration	Reaction time	Residue yield	Residue components			Solution components		
				Glucose	Xylose	Mannose	Glucose	Xylose	Mannose
°C	%	*min*	%[1]	— — %[1] — —			— — %[1] — —		
170	0.4	1.25	74	100	39	40	2	63	55
		2.5	68	98	21	24	2	77	69
		5.0	65	96	13	14	3	77	59
		10.0	62	95	7	20	3	78	64
		20.0	61	94	4	10	7	70	67
	0.8	1.25	67	97	24	29	2	76	75
		2.5	64	97	13	16	3	83	76
		5.0	62	94	8	22	6	79	82
		10.0	59	89	3	8	7	72	61
190	0.4	0.67	68	97	21	13	3	79	69
		1.0	63	96	12	15	3	87	81
		2.0	61	96	5	7	5	83	67
		3.0	60	91	3	11	8	82	72
	0.8	0.67	63	96	10	9	3	83	85
		1.0	60	95	6	9	5	88	75
		1.5	59	92	5	9	8	76	72

[1]% of potential available in original wood.

Table 5.—Calculated and experimental maximum xylose yields from southern red oak (from table 1, sample 1)[1]

Temperature	Acid		Reaction time	Xylan remaining	Maximum xylose yield[3]	Loss
	Added	Effective[2]				
°C	— — % — —		*min*	— — — — % — — — —		
			CALCULATED FROM REMOVAL DATA			
170	0.4	0.29	8.6	8.4	82.7	8.9
	.8	0.69	2.8	12.9	81.3	5.8
	.4	0.29	1.7	6.7	84.6	8.7
190	.8	0.69	0.83	7.5	83.0	9.3
			EXPERIMENTAL YIELDS IN HYDROLYSATES			
170	.4	—	10.0	7	78	15
	.8	—	2.5	13	83	4
190	.4	—	1.0	12	87	1
	.8	—	1.0	6	88	6

[1]Neutralizing capacity = 37 meq/kg.

[2]Calculated using L/S = 4.

[3]Percent of potential.

The above data support the conclusion that maximum xylose yields increase with increasing temperature but are inconsistent as to the effect of varying the acid catalyst concentration. Analysis of data on xylose degradation (Root 1956; Root et al. 1959) leads to the conclusion that the maximum xylose yields should increase with increasing acidity if the pH is above 1.0. Since the use of water (steam) prehydrolysis is frequently proposed, it was of interest to compare the yields from water prehydrolysis, in which the pH is probably greater than 4, to that of dilute acid prehydrolysis. A comparison of the maximum xylose yields was made between water and 0.4% H_2SO_4 as hydrolyzing reagents at 170° C using aspen wood. A large difference was found—the acid solution resulted in a 79% maximum yield of xylose, the water only 61%. Other differences in the systems were also observed and the results published (Springer and Harris 1982). It should be noted that this yield difference should decrease with increasing temperature and is perhaps quite minor at temperatures' above 200° C, which are employed in some processes such as Iotech (1980).

The amount of furfural formed in the prehydrolysis is significant. The xylose degradation can rise to as much as 15% at the point of maximum yield (tables 3 and 5). A model for the complex group of reactions associated with the decomposition of xylose is:

Xylose → Intermediates → Furfural → Decomposition Products

(Intermediates → Decomposition Products; Furfural ← Intermediates)

This model is considerably simplified from actual conditions because there are at least three intermediates, all of which probably react with furfural at different rates. Also, the decomposition of furfural, reacting alone in solution, is not first order, and it is likely that its decomposition products participate in its degradation. Nevertheless, this mechanistic scheme formed the basis for satisfactorily correlating the extensive set of data gathered by Root and coworkers (1956; 1959). The manner in which the correlation can be applied to calculate the production of furfural during prehydrolysis is described in Appendix B. It is assumed that the xylan is hydrolyzed to monomeric xylose, which reacts to form furfural in the yields calculated by Root's correlation.

An important point that can be deduced from the above mechanism is that furfural is produced in high yields during prehydrolysis. In the early stages of the reaction, the concentration of furfural is low, and its loss rate is low since this depends on the furfural concentration. The yield of furfural based on xylose reacted is high, beginning at 100%, and decreases as the amount of xylose reacted increases. Quantitative aspects of this are discussed later.

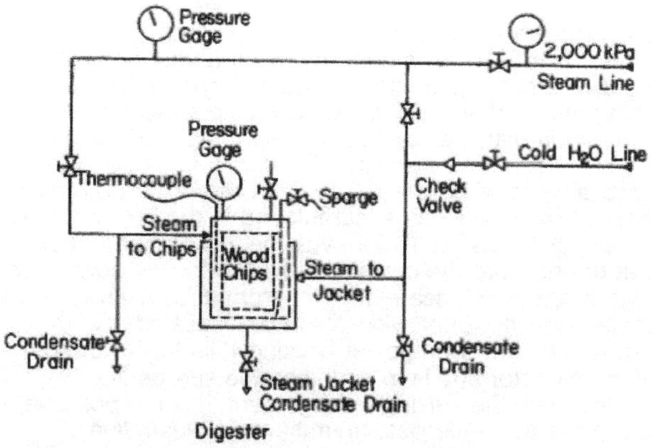

Figure 4.—Schematic diagram of prehydrolysis equipment. (ML84 5466)

Steaming Acid-Impregnated Chips Digester.—Studies on the direct steam prehydrolysis of acid-impregnated chips have been carried out in a small jacketed digester. The internal cavity, 220 mm in diameter, has a slightly dished bottom; maximum depth is 270 mm; its volume is approximately 0.0093 m^3. The closure is a flat, bolt-on cover drilled to accommodate a pressure gage, thermocouple, pressure relief valve, and a line by which condensate collecting in the bottom of the cavity can be removed. Steam admitted into the digester at the midpoint of the sidewall impinges on a small baffle, which prevents the incoming steam from impinging on the chip bed. A schematic of the steam piping is shown in figure 4.

Conditions in the laboratory digester are intended to simulate, as nearly as possible, the conditions in a continuous digester with direct steam heating. Moist acid-impregnated chips are heated to the desired temperature with saturated steam. Some of the resulting condensate is perhaps absorbed by the chips, and some flows down over the chips below. The movement and location of liquid in the bed is important because it affects the catalyst concentration. The cold inner surface of the digester is also a source of condensate. In this small-scale study, the digester-to-charge mass ratio is much greater than it would be in a commercial unit.

To gather meaningful data, it was necessary to separate the digester condensate from the chip condensate. This was done by using a chip container (fig. 5)—a can with a closed bottom, most of its wail being made of screening (A). The chip space is 180 mm in diameter by 150 mm high and holds about 0.5 kg of 9.5-mm oak chips. The can is equipped with a removable false bottom (B) to drain the chips and hold them above the liquid accumulating in the bottom. The lower portion has sufficient volume to hold the liquor that drains from the chips. The lid (C) for the container diverts any condensate dripping from above. The condensate accompanying the incoming steam and that resulting from digester heating falls to the bottom of the digester and is removed by the sparge line (fig. 4). With the valving arrangement, it is not possible to remove all condensate from the main steam line without admitting steam to either the inside or the jacket of the digester. A large amount of condensate accumulates in the steam line when the equipment is idle. This is discharged by opening the valve to the jacket before admitting steam to the chips. This results in a preheating (indirectly) of the chips for about 3 minutes, with the temperature in the vapor space rising to 100°C. Following the release of this accumulated condensate, steam is admitted directly into the inner space; this instant is taken as time zero.

Hydrolyis of impregnated chips.—Digester studies were undertaken with the purpose of relating data gathered in ampoules to the conditions of direct steam heating. in the ampoule procedure, thin wood slices (0.2 mm) contained in seated glass tubes (5 mm) were heated in an oil bath. The major differences between the ampoule and digester cooks are (1) the rate of temperature rise, (2) the changing acid concentration, both in time and space, and (3) the varying US ratio. The preparation of the charge to the digester also has an effect on the rates and yields obtained. The distribution of the acid solution throughout the chips at the time of charging is an important factor; it will depend on the method of impregnation-and the chip size. The chip size will also affect the heatup time and the distribution and concentration of acid during the cook.

in the first runs (1-18), the intent was to study the effect of the condensing steam while minimizing all the other effects. The use of 9.5-mm chips, which have an estimated heatup time of less than 1 minute, minimized the effect of varying temperature. An attempt was made to de-ash the wood to eliminate the effect of the ash constituents and to uniformly impregnate the chips with the acid solution. it was found that the ash could not be completely removed prior to cooking, and, in fact, a significant portion survived the prehydrolysis.

Figure 5.—Chip container for prehydrolysis studies.

Calculations, based on ampoule data, indicated that near-optimum xylose yields would be obtained by reacting at 170°C and a pH, at reaction conditions, of 1.45. This pH corresponds to an effective catalytic acid concentration of 0.343% H_2SO_4. The charge to the digester had an US of ~1.24; the condensate resulting from heating brought this up to about 1.7. Thus, if complete mixing occurred, the concentration of acid at cooking temperature would be 1.24/1.7 = 0.729 times that of the charge concentration. The desired concentration of 0.343% H_2SO_4 would be obtained by adjusting the charge concentration to 0.343/0.729 = 0.47%.

The ash effect was minimized by using the following thorough impregnation procedure to prepare the chips for prehydrolysis: 628 g of 9.5-mm chips (427 g ovendry (OD) wood) were submerged in 6 L of 0.695% H_2SO_4 and vacuum was drawn until no further gas escape was detectable, at which time pressure was restored to atmospheric. The chips were kept submerged to equilibrate for 3 days, at which time the solution was decanted and replaced with a fresh 0.47% acid solution. This was repeated daily for 3 days. Total time in the acid solution was 8 days. The neutralizing capacity of the chips, as measured by ash titration, was reduced by this procedure from 117 to 82 meq/kg. Surprisingly, the reduction was only 56%. Somewhat similar results were found In removing ash from the small disks used in ampoule work where 70% was removed. Grinding to 40 mesh prior to soaking in acid resulted in a 90% reduction in the neutralizing capacity.

Although there were some minor differences in the procedures used and in the data collected, runs 1-18 were handled according to the above description. Southern red oak (table 1, sample 3) 9.5-mm never-dried chips were used. They were impregnated and equilibrated finally with 0.47% H_2SO_4 solution as above. The charge to the digester contained 427 g OD wood (22.2 meq of neutralizing capacity) and approximately 520 g of 0.47% H_2SO_4 solution (25.2 mM $SO_4^=$). The charge was placed in the digester in the container, the digester cover bolted down, and condensate in the steam line drained through the jacket preheating the vapor space to 100°C. At this point (time zero), steam was introduced directly into the vapor space, bringing the temperature to 170°C within 5 seconds. After cooking for a preselected time interval (6-21 min), steam was turned off and cooling water introduced into the steam jacket and also hosed over the digester cover. Two to three minutes were required to lower the internal pressure to below atmospheric, at which point the digester cover was unbolted, and the chip container with its contents placed in a polyethylene bag, and tied to prevent the loss of vapor. During the course of hydrolysis, it was necessary to continuously remove the condensate accumulating in the bottom of the digester. This is condensate resulting from the heating of the digester and, as previously described, was not in contact with the chips. However, volatile components such as furfural and acetic acid were present in this condensate, and so it was collected and assayed. Tables 6, 7, 8, and 9 present a compilation of the data for the various runs including 1-18. The effects of the various phenomena on the hydrolysis rate are interpreted here.

Liquid movement and its effect on acidity.- Comparison of the results of direct steam cooking with those obtained in ampoule studies must be made at the same temperature, time, and catalytic acid concentration. The latter was very difficult to establish, for it depended on many factors. The underlying principle used here was to determine the quantities of cations, $SO_4^=$, and water contained in the solution contacting the solids, and from these calculate the H^+ concentration. Throughout this discussion, the term "residue" refers to the cooked chips and the term "extract" refers to the liquid retained by the "residue." The liquid that drains from the chips and is caught below is called the "free liquor." To determine the effective catalyst concentration, it was necessary to know the amount of cations in the wood charge that were neutralized by the acid, and the distribution of water, cations, and $SO_4^=$ between the free liquor and extract.

Water movement in the digester can be calculated (table 7). Measured values were: water charged, water in the extract after discharge, and water in the free liquor after discharge (col. 3, 7, and 8). Knowing the heat capacity of the charge and the temperature rise and assuming heating occurred by the condensation of saturated steam, the total water at cooking temperature (col. 8) was calculated. Assuming that cooling on discharge occurred by adiabatic vaporization, the total water at discharge (col. 9) was obtained. In a similar manner, from the measurements of the mass and water content of the extract and free liquor after discharge, the water content of these two components at reaction temperature was calculated (col. 4 and 5). The data are irregular; this is apparent if one sums the water in the extract and free liquor (col. 7 and 8) and compares it to the calculated total water (col. 9). However, table 7 does show that the free liquor accumulated very early in the cook. Notice that there was little, if any, increase in the amount of free liquor as the hydrolysis time was extended. This means that there was very little liquid dripping from the chip bed after the initial heatup interval-that is, after the onset of hydrolysis. This is substantiated by the data in table 8, which indicate that 4-9% of the solids solubilized appeared in the free liquor. Runs 32-34 were designed to determine how rapidly the free liquor. accumulated; they show that, at the end of the heatup time, the accumulation of free liquor was essentially complete.

The heating of the chips leads to a reduction in the amount of water retained by the chips (table 7: compare cols. 3 and 4). In the case in which the chips were carefully impregnated, about 30% of the $SO_4^=$ was transferred to the free liquor and, when only a 5-minute impregnation was used, the $SO_4^=$ transferred increased to 40% (table 8). As discussed previously, this transfer of $SO_4^=$ must have occurred during the heatup period. Consideration of these facts led to the conclusion that liquid was forced from the interior of the wood and carried by condensate to the free liquor. A plausible explanation for the driving force that moved this liquid from the interior of the chip is that residual air was present even after vacuum impregnation. Water in the interior of the chip exerted a partial pressure equal to its vapor pressure, which was the same pressure as that at the exterior. The presence of air increased the total pressure in the interior since its partial pressure would be added to that of the water. Thus, to establish pressure equilibrium, the air must have been expelled from the chips, and in its movement to the chip surface, it forced liquid ahead of it.

Table 6.–Distribution of carbohydrate components in digester runs[1]

Run number	Reaction time	Residue yield	Recovery in residue			Recovery in liquids			Water in charge	Residue out		Free liquor out
			Glucan	Xylan	Uronic	Glucan	Xylan	Uronic		Ovendry weight	Liquid	
	min	— — — — — — — — —	*% of original component*	— — — — — — — — —					— — — — —	*kg/100 kg OD wood*	— — — —	
					EXTENSIVELY TREATED CHIPS[2]							
1	15	67.0	100.8	8.3	6.4	5.4	83.1	21.0	135	67.0	105	59.6
2	18	64.7	97.9	5.8	5.4	5.6	83.1	16.5	134	64.7	106	75.4
3	21	65.6	98.5	5.6	4.8	6.1	80.0	12.0	134	65.6	105	
4	9	64.4	95.5	8.5	11.1	3.9	84.5	62.0	116	64.4	100	58.2
5	12	63.2	92.3	7.1	8.2	4.5	82.6	50.9	116	63.2	99	67.2
6	15	62.8	92.1	6.5	6.4	4.7	81.2	27.9	116	62.8	95	66.4
7	6	66.1	94.8	12.4	16.2	3.6	83.7	74.9	128	66.1	102	63.5
8	9	63.6	92.2	8.7	11.1	3.6	81.4	64.9	128	63.6	98	76.4
9	12	65.9	97.8	8.2	8.2	5.7	83.2	53.5	128	65.9	99	65.5
10	6	68.2	101.6	11.4	7.3	4.2	88.4	92.5	118	68.2	109	45.0
11	18	64.2	96.7	5.4	2.4	5.5	83.7	38.1	118	64.2	101	51.1
12	12	64.5	96.2	6.5	8.2	4.7	80.3	53.2	118	64.5	105	53.3
13	12	64.6	92.9	6.8	8.2	4.4	79.5	50.8	120	64.6	108	54.5
14	18	63.9	93.6	6.8	5.4	5.7	80.6	31.0	120	63.9	108	54.3
15	21	63.2	92.6	5.5	4.8	6.3	80.4	24.0	120	63.2	108	67.7
16	8	64.8	97.2	12.1	15.2	3.3	83.6	77.3	123	64.8	110	72.0
17	9	64.3	91.7	10.5	10.8	4.0	84.1	66.7	123	64.3	107	55.2
18	15	63.2	92.4	7.2	6.4	6.0	81.4	42.0	123	63.2	108	67.4
19	6	64.0	93.4	9.5	11.3	5.1	84.6	69.4	127	64.0	111	49.7
20	9	63.0	91.6	6.5	7.7	6.3	82.2	54.8	127	63.0	112	58.8
21	12	62.2	92.7	5.4	5.9	5.9	79.5	42.8	127	62.2	111	52.6
26	4	65.6	95.9	11.2	15.6	3.1	85.6	80.1	133	65.6	120	51.7
27	6	66.4	96.5	9.7	11.7	5.2	86.9	75.7	133	66.4	120	61.7
28	9	63.9	95.0	5.2	7.1	5.7	80.2	53.9	133	63.9	113	63.1
29	4	68.7	98.6	18.1	25.9	2.3	82.0	75.0	130	68.7	120	51.6
30	6	66.9	94.6	13.1	17.1	3.4	84.0	75.8	130	66.9	118	84.8
31	9	65.0	93.6	11.1	11.3	3.6	84.1	64.5	130	65.0	117	50.6
					RAPIDLY TREATED CHIPS[3]							
35	6	64.1	95.6	9.8	11.4	5.7	80.9	66.8	110	64.1	100	52.4
36	9	61.9	91.8	5.4	7.4	7.1	74.8	47.9	111	61.9	99	55.9
37	12	62.0	91.1	6.8	5.9	8.8	70.8	37.4	109	62.0	96	59.1
38	6	66.6	94.1	12.2	14.0	5.1	78.9	72.0	110	66.6	99	51.1
39	9	64.8	95.4	8.9	11.8	6.2	78.3	56.2	113	64.8	103	60.3
40	12	65.3	93.9	11.0	8.7	6.8	76.1	44.9	115	65.3	97	57.9
41	6	67.2	96.1	8.4	11.7	5.6	79.9	67.9	118	67.2	102	58.8
42	9	66.3	96.3	6.4	9.9	6.7	74.6	51.6	117	66.3	99	55.0
43	12	65.4	93.7	8.0	8.5	7.0	72.0	41.9	116	65.4	98	53.5

[1]All runs at 170° C with 9.5-mm southern red oak chips(from table 1, sample 3).

[2]Chips were partially de-ashed, carefully impregnated, and equilibrated with acid solution for 6 days.

[3]Chips were given a 5-minute vacuum impregnation and cooked immediately.

Table 7.—Water distribution in digester runs[1]

Run number	Reaction time	Water charged	Water before discharge			Water after discharge		
			Extract	Free liquor	Total	Extract	Free liquor	Total
1	2	3	4	5	6	7	8	9
	min	*kg/100 kg OD wood*						
			EXTENSIVELY TREATED CHIPS[2]					
1	15	135.2	101.3	73.0	187.4	73.7	58.0	142.1
2	18	134.5	100.0	92.7	186.5	72.6	73.7	141.3
4	9	116.2	91.8	71.0	162.7	66.1	56.4	122.4
5	12	116.0	88.9	82.2	162.4	63.8	65.3	122.1
6	15	115.8	84.2	81.3	162.0	60.1	64.6	121.9
7	6	128.2	96.0	78.2	178.3	69.4	62.1	134.8
8	9	128.2	88.6	94.1	178.3	63.5	74.8	134.8
9	12	127.9	92.9	80.2	177.9	66.9	63.7	134.5
10	6	117.6	107.6	54.5	164.5	78.7	43.3	123.8
11	18	117.6	92.8	61.7	164.5	66.9	49.0	123.8
12	12	117.6	98.8	64.9	164.5	71.7	51.5	123.8
13	12	119.5	102.4	66.4	166.9	74.5	52.7	125.8
14	18	120.0	100.9	66.1	167.6	73.3	52.5	126.3
15	21	120.2	100.7	82.8	167.9	73.1	65.8	126.5
16	6	123.3	104.7	88.4	171.8	76.3	70.3	129.7
17	19	122.8	100.6	67.3	171.2	73.1	53.5	129.2
18	15	123.5	101.0	82.5	172.1	73.4	65.6	129.9
19	6	127.2	104.8	60.6	176.9	76.4	48.1	133.7
20	9	127.2	105.4	71.5	176.9	76.9	56.8	133.7
21	12	127.2	103.3	63.6	176.9	75.3	50.5	133.7
22	6	184.2	176.7	105.0	251.4	132.2	82.7	191.0[3]
23	6	168.8	161.2	100.4	231.4	120.0	79.1	175.2[4]
24	6	184.2	215.4	111.6	251.4	161.2	87.3	189.3[3]
25	6	168.8	163.1	95.7	231.4	122.5	75.8	176.7[4]
26	4	133.1	119.6	62.7	184.7	87.6	49.5	139.0
27	6	133.1	122.5	74.5	184.7	89.7	58.7	138.7
28	9	133.1	111.2	75.9	184.7	80.8	59.7	138.5
29	4	130.2	124.2	63.3	180.9	90.7	49.8	135.4
30	6	130.2	119.3	105.0	180.9	86.7	82.5	135.0
31	9	130.2	118.5	61.7	180.9	84.6	47.8	132.9
			RAPIDLY TREATED CHIPS[6]					
32	1	109.9	102.7	67.8	154.4	73.9	53.6	114.7
33	2	112.7	99.1	66.4	158.1	70.8	52.4	117.4
34	4	113.2	94.1	72.3	158.7	66.2	56.6	116.8
35	6	109.7	93.5	64.5	154.1	66.3	50.6	113.9
36	9	110.9	88.3	69.1	155.6	62.2	54.3	115.1
37	12	109.0	84.3	72.8	153.2	59.5	57.5	113.8
38	6	109.9	96.0	62.2	154.4	68.2	48.7	114.0
39	9	113.0	98.4	72.7	158.4	70.7	57.3	117.9
40	12	114.8	92.7	71.0	160.8	65.1	55.3	118.2
41	6	118.3	99.1	70.5	165.4	72.1	56.0	124.9
42	9	117.4	93.6	66.1	164.2	67.6	52.4	123.6
43	12	115.5	94.2	65.4	161.7	66.0	50.9	118.6

[1]All runs at 170° C with 9.5-mm southern red oak chips (from table 1, sample 3), except as noted.

[2]Chips were partially de-ashed, carefully impregnated, and equilibrated with acid solution for 6 days.

[3]Aspen.

[4]Birch.

[5]Chips were given a 5-minute vacuum impregnation and cooked immediately.

Table 8.—Component distribution in digester runs[1]

Run number	Reaction time	Soluble solids		Sulfate		Cations		Calculated pH
		Total	In free liquor	Charged	In free liquor	Charged	In residue	
	min	*kg*[2]	%[3]	*g mols*[2]	%[4]	eq[2]		
			EXTENSIVELY TREATED CHIPS[5]					
1	15	33.0	—	6.5	27.0	5.2	—	1.67
2	18	35.3	—	6.5	27.8	5.2	—	1.68
3	21	34.4	—	6.5	27.3	5.2	—	1.68
4	9	35.6	—	5.6	31.4	5.2	—	1.85
5	12	36.8	—	5.5	34.0	5.2	—	1.88
6	15	37.2	—	5.5	32.1	5.2	—	1.81
7	6	33.9	4.1	6.2	—	5.2	—	1.68
8	9	36.4	4.3	6.2	—	5.2	—	1.64
9	12	34.1	5.1	6.2	—	5.2	—	1.67
10	6	31.8	5.5	5.6	—	5.2	2.5	1.80
11	18	35.8	6.0	5.6	—	5.2	2.1	1.78
12	12	35.5	—	5.6	—	5.2	—	1.81
13	12	35.4	—	5.7	—	5.2	—	1.81
14	18	36.1	—	5.7	—	5.2	—	1.80
15	21	36.8	—	5.7	—	5.2	—	1.80
16	6	35.2	—	5.9	—	5.2	—	1.79
17	9	35.7	—	5.9	—	5.2	—	1.77
18	15	36.8	—	5.9	—	5.2	—	1.77
19	6	36.0	4.4	8.1	29.9	3.0	1.8	1.37
20	9	37.0	5.5	8.1	33.2	3.0	1.5	1.43
21	12	37.8	5.5	8.1	31.8	3.0	1.0	1.46
26	26	34.4	6.4	8.8	28.1	3.8	2.9	1.30
27	27	33.6	8.9	8.9	30.4	3.8	2.5	1.35
28	9	36.1	9.4	8.9	30.4	3.8	1.9	1.36
29	4	31.3	5.7	5.6	28.7	3.7	2.2	1.60
30	6	33.1	6.7	5.6	28.5	3.7	2.5	1.55
31	9	35.0	8.0	5.5	30.3	3.7	2.1	1.63
			RAPIDLY TREATED CHIP[6]					
32	1	—	—	14.7	38.5	11.7	—	1.46
33	2	—	—	15.5	38.0	11.7	—	1.38
34	4	—	—	15.6	43.8	11.7	—	1.45
35	6	35.9	5.0	16.9	40.4	11.7	2.4	1.30
36	9	38.1	4.3	17.3	41.2	11.7	1.7	1.31
37	12	38.0	4.2	16.7	45.6	11.7	1.6	1.42
38	6	33.4	7.2	14.4	38.5	11.8	2.7	1.41
39	9	35.2	8.7	14.7	39.9	11.8	2.3	1.46
40	12	34.7	7.3	13.8	43.3	11.8	2.0	1.61
41	6	32.8	8.6	17.5	38.0	11.8	2.8	1.23
42	9	33.7	7.7	18.0	38.3	11.8	2.1	1.21
43	12	34.6	7.6	15.6	40.6	11.8	2.4	1.38

[1]All, runs at 170° C with 9.5-mm southern red oak chips (from table 1, sample 3).

[2]Basis: 100 kg OD wood charged.

[3]Based on total soluble solids.

[4]Based on sulfate charged.

[5]Chips were partially de-ashed, carefully impregnated, and equilibrated with acid solution for 6 days.

[6]Chips were given a 5-minute vacuum impregnation and-cooked immediately.

Table 9.—Furfural yields from the prehydrolysis of southern red oak

Run number	Reaction time	Xylose reacted	Uronic reacted	Potential furfural		Yields	
				Xylose	Uronic	On xylose	On wood
	min	% [1]	% [1]	% [2]	% [2]	%	%
29	4	—	—	—	—	—	0.27
30	6	2.9	7.1	.39	.13	131.03	.51
26	4	3.2	4.3	.43	.08	127.19	.55
27	6	3.4	12.6	.46	.23	193.24	.89
7	6	3.9	8.9	.52	.16	111.28	.58
16	6	4.3	7.5	.58	.13	69.07	.40
31	9	4.8	24.2	.65	.43	132.39	.86
17	9	5.4	22.5	.72	.40	100.56	.73
19	6	5.9	19.3	.80	.35	99.15	.79
4	9	7.0	26.9	.94	.48	92.29	.86
9	12	8.6	38.3	1.15	.68	105.81	1.22
1	15	8.6	72.6	1.15	1.29	122.09	1.41
38	6	8.9	14.0	1.19	.25	87.19	1.04
35	6	9.3	21.8	1.24	.39	109.68	1.37
8	9	9.9	24.0	1.32	.43	74.75	.99
5	12	10.3	40.9	1.38	.73	91.26	1.26
2	18	11.1	78.1	1.49	1.38	103.60	1.54
20	9	11.3	37.5	1.53	.67	74.78	1.14
18	15	11.4	51.6	1.53	.91	92.98	1.42
41	6	11.7	20.4	1.57	.36	74.10	1.16
6	15	12.3	65.7	1.65	1.16	100.61	1.66
14	18	12.6	63.6	1.69	1.13	94.44	1.59
39	9	12.8	32.0	1.71	.57	83.59	1.43
40	12	12.9	46.4	1.73	.82	108.53	1.87
12	12	13.2	38.6	1.77	.68	59.85	1.06
13	12	13.7	41.0	1.83	.73	60.58	1.11
15	21	14.1	71.2	1.89	1.26	102.13	1.93
3	21	14.4	83.2	1.93	1.48	70.83	1.37
28	9	14.6	39.0	1.98	.70	76.03	1.50
21	12	15.1	51.3	2.04	.92	76.82	1.57
28	9	19.0	38.5	2.54	.68	75.79	1.93
30	9	19.8	44.7	2.65	.79	74.24	1.97
43	12	20.0	49.6	2.68	.88	81.50	2.18
37	12	22.4	56.7	3.00	1.01	82.00	2.46

[1] % of original component

[2] % of wood.

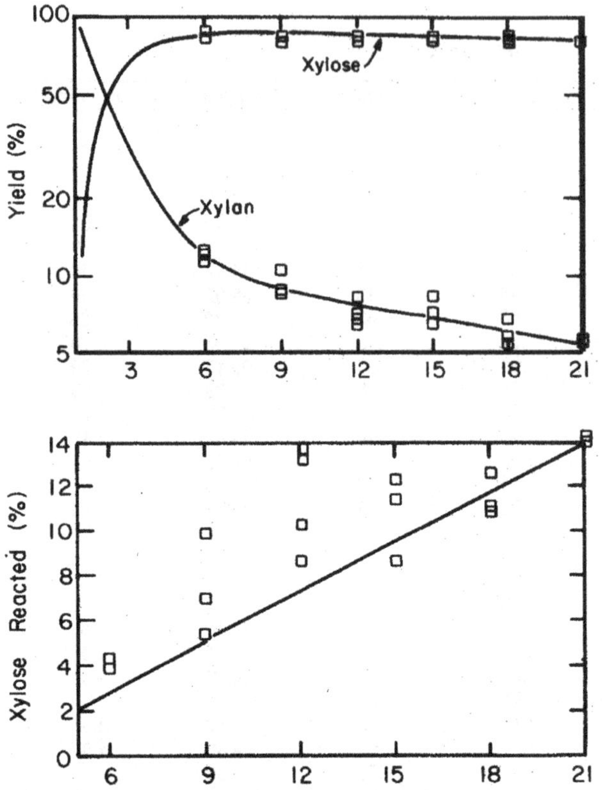

Figure 6.—Yields of xylan and xylose over time (upper), and xylose reacted over time (lower) in prehydrolysis of southern red oak, runs 1-18, 170° C, pH = 1.77. Solid lines are calculated. (ML84 5473)

This movement of water should depend on several factors, including the morphology of the wood, the method of impregnation, and perhaps the liberation of CO,. These possibilities were the motivation for runs 22-25 (table 7) using aspen and birch. The same water movement was apparent, but its magnitude was not as great. Neither of these species attain the low L/S of the charge that is possible with oak, with the result that much greater quantities of water were present in both the extract and free liquor. Although not definitive, the data in table 7 seem to indicate that the amount of extract was decreased by the more rapid impregnation supporting the analysis above.

Only In some runs were measurements made on the $SO_4^{=}$ content of the free liquor and on the amount of unreacted cations in the residue (table 8). Where these were not measured, assumed values were used to calculate the acidity of the extract. For funs 1-18, it was assumed that 30% of the sulfate charged was in the free liquor and that the residue contained 2.2 eq of cations/100 kg OD wood charged. In no case, for the thoroughly impregnated chips, were measurements taken of the amount of cations in the free liquor. It was assumed that the value was 15% of the cations solubilized, which is the average of measurements made for other runs. This completed the information required to calculate the acidity using the procedure in Appendix A. The results are given in the last column of table 8. The average pH of 1.77 for runs 1-18 corresponded to a sulfuric acid concentration of about 0.17%; comparing this to the strength of the impregnating solution (0.475%) indicates how greatly these factors affect the acid strength.

Data correlation.—The xylose yield data of table 6 pertaining to runs 1-18 is compared with calculated yields in figure 6. In table 6, the values reported for the various carbohydrates recovered In solution include all polymeric forms as well as the free monomer. The lower curve of xylan removal (fig. 6 upper), correlated as previousiy described, is the sum of two exponential terms, specifically:

XR = xylan remaining in residue, %

$$= 88.5 \exp(-0.713t) + 11.5 \exp(-0.0378t) \qquad (6)$$

where t = time (min).

This analytical expression for the liberation of xylose, together with the acid concentration of 0.17% H_2SO_4 (pH = 1.77) and the US, was sufficient to calculate the yield for xylose in solution (fig. 6 upper). The calculated maximum yield was 86.2% occurring at 8 minutes; the experimental data indicate a maximum yield of 83.8% at 6 minutes, with the yield at 9 minutes being 83.5%. The logarithmic scale obscures the fit of the data; a much more critical presentation is made in figure 6 lower. The data of table 6 contain, implicitly, the percentage of xylose reacted; it is the sum of the xylan in the residue and xylan in solution subtracted from 100. These values are compared to the calculated amount of xylose reacted (the curve in fig. 6 lower). All the information in figure 6 lower is contained in figure 6 upper, but the lower portion is a more stringent test of the data.

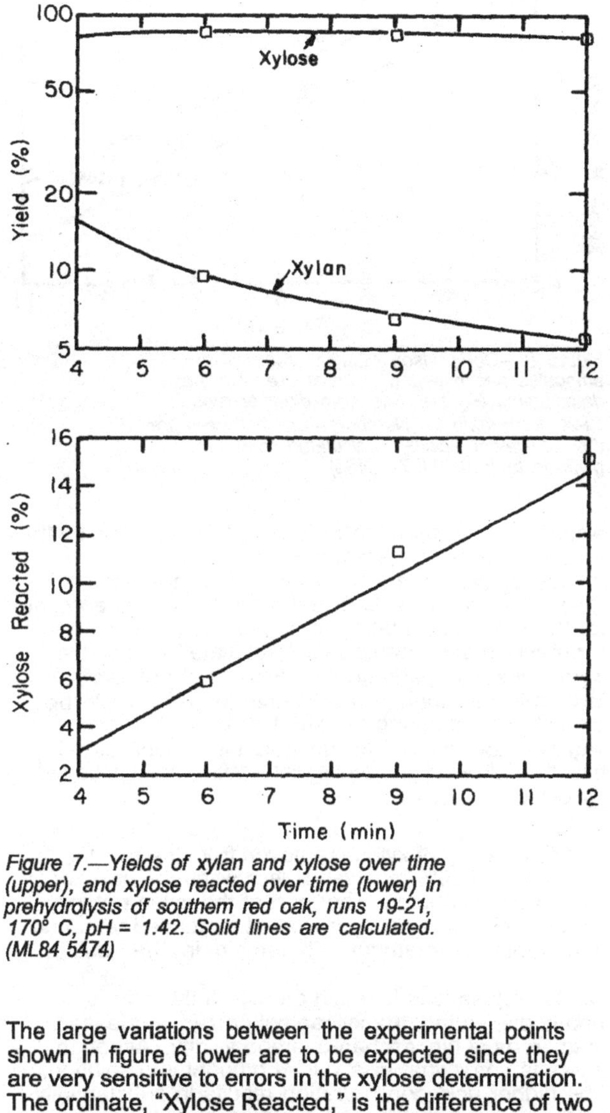

Figure 7.—Yields of xylan and xylose over time (upper), and xylose reacted over time (lower) in prehydrolysis of southern red oak, runs 19-21, 170° C, pH = 1.42. Solid lines are calculated. (ML84 5474)

The large variations between the experimental points shown in figure 6 lower are to be expected since they are very sensitive to errors in the xylose determination. The ordinate, "Xylose Reacted," is the difference of two large numbers, one of which is dependent on three separate analyses for xylose. Most of the experimental values fall above the calculated quantity of xylose decomposition-that is, xylose reacted. The data would be best represented by assuming an acid concentration of 0.2% rather than the value of 0.17%, which was used. This difference was judged to be within the limits of error for this group of early runs, and it was concluded that the prehydrolysis is adequately described by the above analysis.

The early runs (1-18) had an acidity considerably lower than the desired pH of 1.45. Runs 19-21 with fully impregnated oak chips were closer to the desired pH. The method of treating these chips was somewhat different. The neutralizing capacity was lowered to 30 meq/kg OD wood by first treating the chips with hydrochloric acid and washing with distilled water. They were then long-term vacuum impregnated as before but using a solution of 0.72% H_2SO_4. Measurements made on the sulfate content of the extract and free liquor after hydrolysis showed that the composition of the solution associated with the wood charge must have been 0.63% H_2SO_4. It is quite probable that the expected uptake of the impregnating acid was not reached because of the presence of non-solvent water in the wood-that is, water not available for equilibrating with acid (Sookne and Harris 1940). The unneutralized cations remaining in each of the residues were also measured, and, as shown in table 8, their values decreased significantly with time. This resulted in a changing acidity during the cook-at 6 minutes the pH was 1.37, corresponding to 0.417%. H_2SO_4; by 12 minutes the pH rose to 1.46 or an equivalent acid concentration of 0.338%. The average effective acid concentration over the interval 6 to 12 minutes was 0.38% (pH = 1.42).

The plots for runs 19-21 (fig. 7) are similar to those for runs 1-18. in this case, the removal curve equation is:

$$XR = 87.0 \exp(-0.914t) + 13.0 \exp(-0.0815t) \quad (7)$$

The maximum calculated xylose yield is 84.5% occurring at 6 minutes, whereas the measured maximum is 84.6% at 6 minutes. The experimental values of xylose reacted agree very well with those calculated (fig. 7 lower). Similar analysis of the data from runs 26-28 and 29-31 confirm the fact that xylose yields may be satisfactorily calculated from the experimentally determined removal curve.

One may compare the data obtained from direct steam heating in a digester to that from ampoule experiments. At 170° C the maximum xylose yields obtained in ampoules using 0.4% H_2SO_4 (0.29% effective acid) and 0.8% (0.69%) were 78% and 83%, respectively, indicating a modest increase in yield with acidity (table 4). The experimental maxima obtained in the digester were 83.8% with 0.17% H_2SO_4 (runs 1-18) and 84.6% with 0.38% H_2SO_4 (runs 19-21). The ampoule and digester studies used different wood samples, employed different analytical procedures, and were done by different personnel. The reported higher yields from the digester are probably not significant.

It was concluded that the xylose yields obtained by direct steaming were the same as obtained in ampoules when compared at the same acidity. However, the two procedures differed in their rates of xylan removal (fig 8). it appears that, in direct steam heating, the xylan removal rate is lower than what would be predicted from the acidity of the extract. This is perhaps due to nonuniform distribution of acid resulting either from incomplete impregnation prior to heating or from the movements of water and acid during the hydrolysis.

Prehydrolysis of rapidly impregnated chips.—In contrast to previous runs in which chips were partially de-ashed and carefully impregnated, runs 32-43 in tables 7, 8, and 9 used a rapid and more practical procedure. The 9.5-mm chips (table 1, sample 3, 32% moisture) were submerged in a sulfuric acid solution, vacuum was drawn with a water aspirator for 5 minutes, and the pressure was restored to atmospheric. The liquid was then drained from the chips; they were shaken free of liquid, weighed to obtain the solution pickup, and placed in the digester to be prehydrolyzed immediately.

Runs 32-37 were impregnated in a 2.5% sulfuric acid solution, 38-40 with 2.0%, and 41-43 with 2.25%, all for 5 min.

It was found that the quantity of sulfuric acid associated with the chips could not be calculated from the amount of solution pickup and the strength of the impregnating solution; it was determined experimentally by analyzing for $SO_4^{=}$ in the hydrolysate liquors. The sulfate pickup on impregnation is 7-22% greater than that calculated on the basis of solution pickup. Since a similar phenomenon is observed when chips are impregnated with base, it is assumed that the additional sulfate pickup is due to diffusion from the bulk liquid to the solution within the chip structure. A concentration gradient exists because the solution drawn into the chip will be diluted by the original moisture in the wood.

The rapidly impregnated chips have substantially less water in the charge and significantly lower quantities of extract than the carefully impregnated chips (table 7). Much larger fractions of the charged acid appear in the free liquor when chips are rapidly impregnated (table 8).

For some runs, measurements were made of the Ca^{++} and K^+ transferred to the free liquor; on the average, 15.2 and 14.3% of the original sum of Ca^{++} and K^+ (equivalents) in the wood were found in the free liquor. For those runs using rapid impregnation, it was assumed that the charged chips contained all of their original ash components. It was further assumed for the calculations of table 8 that 15% of the cations charged were transferred to the free liquor during hydrolysis.

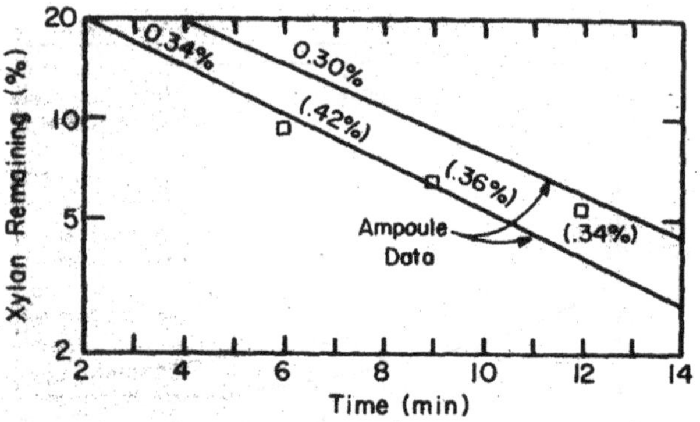

Figure 8.—Comparison of xylan removal in ampoules and digester. Curves are ampolue data; points (•) are data from digester runs 19-21 from table 6. Numbers in parentheses are the equivalent acidities calculated from the pH given in table 8. (ML84 5485)

Analysis of the data for the rapidly treated chips by the procedure used for previous runs is quite unsatisfactory. The calculated acid concentrations, reported as pH in table 8, are much too low to account for the rapid degradation of xylose. Measurements of furfural formation (table 9) substantiate the reported xylose losses. Evidently the acid is not uniformly distributed throughout the substrate. This would be expected considering the short impregnation and equilibration times. The unequal distribution of acid has the following consequences, which are supported by data:

1. The quantity of xylose removed from the residue (in the range of 90% removal) is less when the acid is not uniformly distributed. That part of the xylan not in contact with acid will hydrolyze at a reduced rate, and much of this material will remain in the residue.

2. The xylose loss is much greater. If the acid is poorly distributed, the xylose solubilized is released into areas of higher-than-average acidity. Although there is somewhat less xylose brought into solution, the higher acid concentration results in substantially greater losses than found in a system with a uniform distribution.

3. The maximum yield of solubilized xylose is decreased. This is evident from the data of table 6 where a decrease of approximately 5% is indicated. This results from the slower release and more rapid decomposition of the xylose.

4. Furfural production is greater. This results from the increased xylose decomposition. The maximum combined yield of xylose and furfural from the nonuniformly distributed acid medium is only 1-2% less than-that from the system in which the acid is uniformly distributed.

The distribution of acid throughout the reacting substrate Is undoubtedly one of the important factors determining the rate of xylose removal and, consequently, xylose and furfural yields. The wood species, chip size; and method of impregnation all. influence the acid distribution. Detrimental effects due to nonuniformly distributed acid are apparent in the data gathered for 9.5-mm chips. Even greater effects would be expected for larger chips.

Furfural yields.—Since furfural is a degradation product of both xylose and uronic acid, it is unavoidably produced in the prehydrolysis. Experimental yields are shown in the last column of table 9. The extent of the reaction of both xylose and uronic acid is also shown; these values are obtained by summing the recoveries in the extract and residue (table 6) and subtracting from the charge (100%). The relative importance of each as furfural sources if they were converted in stoichiometric yield is indicated by the potential furfural yields given in table 9. it is apparent that uronic acid could contribute appreciably to the furfural yield. However, uronic acid degrades to products other than furfural (Feather 1973), and yields probably do not exceed 30% of stoichiometric. The lack of information on furfural yields from uronic acid makes it difficult to interpret the furfural yields from prehydrolysis. However, projected yields from the reacting xylose can be calculated and compared with the experimental values.

We pointed out earlier that the mechanism of furfural. formation is such that the stoichiometric yield (based on xylose reacted) is initially 100% and decreases as the reaction proceeds (fig. 9). The integrated value is 100 times furfural produced/xylose reacted over the entire reaction period, while the differential value is the same ratio but based on an infinitesimal interval. The differential efficiency assumes negative values after the point of maximum furfural yield (based on charge) is reached.

The integrated yield values in figure 9 were used to construct the curve in figure 10 where the yield is based on wood charged. Many of the experimental points shown in figure 10 fall above the predicted yield from the the xylose alone. There are two reasons why this should be expected: (1) The uronic acid must make some contribution to furfural yield, and (2) much of the furfural escapes from the extract, being also present in the free liquor and condensate, and thus is removed from the reaction site. In general, more than half of the furfural is recovered in the condensate. Consideration of the data in figure 10 leads to the conclusion that the yield of furfural can be estimated from the calculated curve.

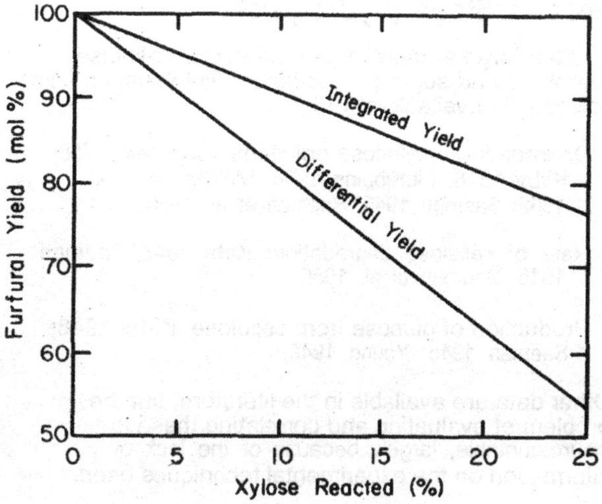

Figure 9.—Calculated efficiency of furfural generation from xylose. Assumed prehydrolysis conditions were 170° C, pH = 1.45 (App. B). (ML84 5486)

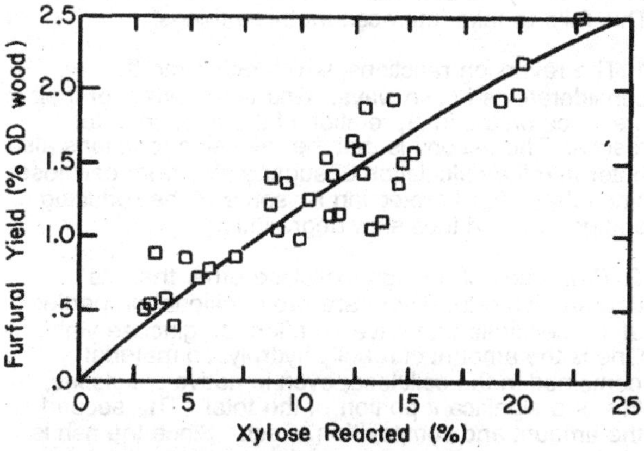

Figure 10.—Furfural yields obtained in digester prehydrolysis. The curve is the yield from xylose (excluding uronic acid) reacted, based on the integrated yield curve of figure 9. (ML84 5487)

Second Stage (Hydrolysis)

A rather large amount of data related to cellulose hydrolysis and sugar production in dilute sulfuric acid solutions is available at FPL:

Degradation of glucose solutions: Dunaway 1950; Kirby 1948; McKibbins 1958; McKibbins et al. 1962; Saeman 1945; Saeman et al. 1950.

Rate of cellulose degradation: Kirby 1948; Saeman 1945; Saeman et al. 1950.

Production of glucose from cellulose: Kirby 1948; Saeman 1945; Young 1949.

Other data are available in the literature, but the problem of evaluating and correlating these is insurmountable, largely because of the lack of information on the experimental techniques used.

In the past, the usual procedure for estimating glucose yield from cellulose hydrolysis was elementary. It was assumed the system could be modeled simply as two consecutive reactions, each proceeding at the rate predicted by the correlations given by Saeman (1945). This fails to take into account three things:

1. The reversion reactions, which enter into the considerations in two ways. One is the effect of their presence on the interpretation of the experimental results. The second is that the reversion reactions also enter into the calculation of sugar yields from cellulose since they afford protection for some of the reducing end groups and thus slow degradation.

2. Properties of the lignocellulose other than its degradation rate. There are two additional properties of the cellulose that have an effect on glucose yield. One is the amount of readily hydrolyzed material contained in the cellulose; even in native celluloses, this is a significant portion of the total. The second is the amount and composition of ash. Since the ash is basic, it lowers the acidity of the reaction mixture; the catalytic hydrogen ion is a function of the concentration and amount of the applied acidic solution and the neutralizing power of the ash. Saeman (1945) recognized this factor but did not make a correction for the effect of ash on the hydrogen ion concentration.

3. The existence of data subsequent to 1944. All the available information should be brought together, evaluated, and fit into a model which could be used to predict yields of all the components produced during cellulose hydrolysis.

There are rather few external factors that affect the yields and rate of the second-stage dilute acid hydrolysis:

1. Temperature--including its variation with time.

2. Concentration of applied acid (H_2SO_4).

3. Liquid-to-solid ratio, L/S.

4. Properties of the lignocellulose.

Individual phenomena that must be known and understood to predict the glucose yield and rate of cellulose hydrolysis are:

1. The production and role of reversion products.

2. Rate of glucose degradation as a function of temperature, catalyst strength, and glucose concentration. The possible effects of degradation products on the rate can probably be ignored.

3. Rate of cellulose degradation as a function of temperature, catalyst concentration, and L/S.

Each of these will be considered separately and then combined into a single model.

Glucose Reversion

Reversion may be defined as the acid-catalyzed formation of glycosides from monosaccharides. The products of the reaction are a complex mixture consisting principally of dimers, oligosaccharides, and internal glycosides (anhydrosugars). These appear in admixture with glucose and other materials from the hydrolysis of cellulose. The proportion of reversion materials in the hydrolysis products is small in dilute sugar solutions but significant at the concentration levels encountered in this study.

Review of data (BeMiller 1965; Kerr 1944; Spriggs 1944; Sroczyński and Boruch 1964) on the composition of glucose reversion products indicated that, by far, the most prominent constituents were [1,6]-linked glucosides, The literature data and out initial experimental work at 180° C indicated that the disascharides—gentiobiose and isomaltose—accounted for most of the lower molecular weight material, with the internally [1,6]-linked glucoside—levoglucosan—constituting less than 30% of the total reversion products. However, further experiments, described below show levoglucosan to be a more prominent product at 230° C. These facts led to the adoption of the reaction scheme shown in figure 11. The glucosides shown contain an aglycone formed by dehydrating the reducing end group of the disaccharide. Hydroxymethylfurfural could reasonably be one such aglycone, in which case the glucoside would be 2-furfuryl-5-methyl-glucoside, a product that has been reported in reversion mixtures (Sroczyński and Boruch 1964). Notice that the rate of dehydration of the disaccharides has been assumed equal to that of glucose (k_1). The model ignores the presence of oligomers with degree of polymerization (DP) greater than 2, which, in highly concentrated glucose solutions, are major components of the reversion material. Using the equilibrium constants reported below it is estimated that at 230° C the higher oligosaccharides (DP>2) in a 200-mg/mL solution of glucose would not exceed 15% of the combined material. The scheme shown in figure 11 is considered adequate to incorporate reversion information into a cellulose hydrolysis model.

The equilibrium constants CD and CL (defined in App. C) were obtained from the experimental data reported (table 10). These data were collected and analyzed by the following procedures. Acidified glucose solutions were sealed into 3-mm glass ampoules and heated for various times in a salt bath. After reaction, the solution was analyzed with glucose oxidase to determine the monomeric glucose. The sample was then fermented to remove most of the monomeric glucose leaving the reversion material. After fermentation of the glucose, the remaining material was analyzed for monomeric glucose and reducing power. The reversion products were then submitted to a mild acidic hydrolysis, which converted the glucosides to glucose, and again analyzed for monomeric glucose and reducing power. Corrections were made for the glucose loss occurring during the mild acidic hydrolysis. The values reported are averages of three to six replicates. A complete report of these data has been given by Minor (1983).

The data of table 10 were handled as described in Appendix C. The method of analyzing the data defines the various terms. Reversion products are those components that liberate glucose on mild acidic hydrolysis and would thus include all glucosides. All nonproducing glucosides are reported as levoglucosan; the apparent levoglucosan content would be increased by the presence of oligosaccharides (DP>2). The reported amount of levoglucosan is also influenced by the value assumed for the reducing power of the disaccharides. The increased quantity of levoglucosan present at the higher temperature was confirmed by direct analysis.

The parameters CD, CL, and k, are sufficient to describe the system of figure 11 if the reversion reactions are at equilibrium (assuming $k_6 >> k_1$). But at what rate is equilibrium attained? Examining the constants CD and CL (table 11), one concludes that levoglucosan is at equilibrium for each of the time intervals, but the disaccharides may not be at equilibrium even after the 36-second time interval. The following is meant to establish the validity of assuming reversion equilibrium during cellulose hydrolysis.

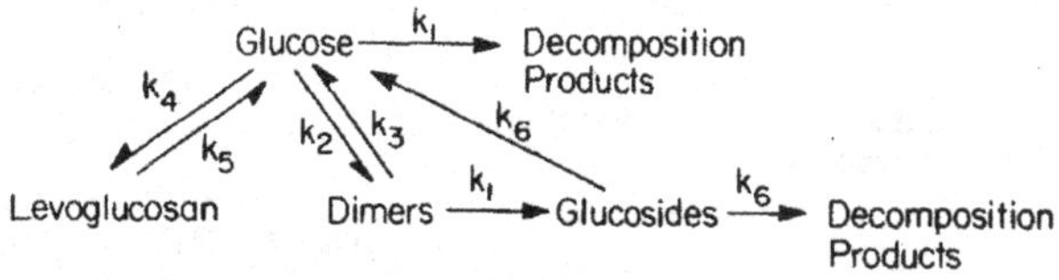

Figure 11.—Model for glucose reversion. (ML84 5476)

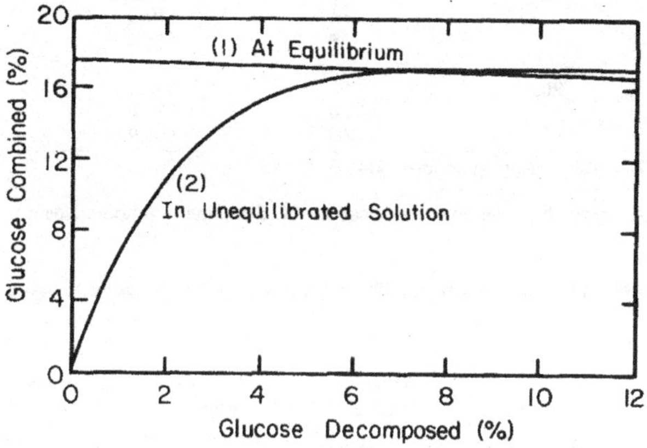

Figure 12.—Attaining reversion equilibrium in a 1M glucose solution. Curve (1) is percentage of total glucose combined at complete equilibrium, and curve (2) is percentage of total glucose combined in unequilibrated solution. (ML84 5490)

The rate of hydrolysis of all glycosides is rapid compared to the rate of degradation of glucose. The hydrolysis rate of isomaltose is approximately 35 times the rate of glucose decomposition, and the ratio is even higher for gentiobiose. By assuming values for the rates of the glucoside hydrolysis relative to glucose decomposition and using the equilibrium constants from table 11, one can relate the appearance of the reversion products to the disappearance of glucose. The result of such a calculation for a 1M glucose solution, using the assumed values of $k_3/k_1 = k_5/k_1 = 35$ (fig. 11), is shown in figure 12. The equilibrium curve shows the amount of combined glucose that would be present if the system were at complete equilibrium; it decreases as glucose decomposes because of the falling glucose concentration. The system, however, does not reach equilibrium immediately. The quantity of combined glucose, disaccharides plus levoglucosan, increases rapidly, finally attaining the equilibrium value when 7.0% of the glucose is decomposed, and thereafter is only slightly higher than the equilibrium value, Thus, once equilibrium is established, it can be assumed to be maintained. During cellulose hydrolysis, although there is a rapid release of a small amount of glucose in the initial phase, the system is in equilibrium at the beginning of the reaction and equilibrium should be closely maintained throughout the entire reaction.

Table 10.—Reversion data for glucose reacted at 230° C and 0.1% H_2SO_4

Initial concentration	Reaction time	Monomeric glucose[1]			Reducing analysis[2]	
		Before fermentation	After fermentation	After hydrolysis	After fermentation	After hydrolysis
1	2	3	4	5	6	7
mg glucose/mL	*sec*			*mg glucose/mL*		
150	14	122.1	1.00	20.74	4.13	22.09
	24	119.9	1.48	21.10	4.17	22.54
	36	115.5	.47	21.27	4.65	23.25
200	14	156.1	1.04	29.84	5.65	32.18
	24	146.8	1.38	30.53	7.00	32.46
	36	144.9	1.35	31.05	7.77	34.06

[1]Analysis using glucose oxidase.

[2]Analysis by Nelson's modification of Somogyi method (Nelson, 1944).

Table 11.—Reversion equilibrium constants for glucose reacted at 230° C and 0.1% H_2SO_4

Initial concentration	Reaction time	Glucose loss	Combined glucose	Reducing power[1]	Equilibrium constants[2]		Fraction levoglucosan
					CD	CL	
mg glucose/mL	*sec*	%	%				
150	14	5.0	14.3	0.153	0.038	0.116	0.69
	24	6.5	14.5	.132	,034	.125	.74
	36	8.4	15.9	.192	.056	.116	.62
200	14	7.0	16.1	.154	,034	.133	.69
	24	11.6	17.0	.187	,047	.129	.63
	36	11.9	17.7	,206	,055	.127	.59

'Reducing power = mols glucose/mol combined glucose.

[2]Equilibrium constants: D = (CD) x G^2; L = (CL) x G.

Glucose Decomposition

An evaluation of the scope of the available experimental data (Dunaway 1950; Kirby 1948; McKibbins 1958; McKibbins et al. 1962; Saeman 1945; Saeman et al. 1950) (table 12) shows that the most extensive are those of McKibbins, which cover the high-temperature and low-acid region of interest here, and also include data for up to 20% glucose concentration. This has led to the decision to use McKibbins' correlation, but with some reservation. There is considerable difference between Kirby's and McKibbins' data, even in the area where they overlap. All the information should be reexamined and recorrelated; new data should be added to reinforce and extend the correlation. However, McKibbins' correlation gives a reasonable estimate of the rate constant at any condition in the range of interest. They are the only available data that include measurements at low acidity.

In all of the above studies, the solution property measured was the reducing power. Time intervals in all tests were of sufficient duration that reversion equilibrium was established; that is, data points fell on the straight-line portion of the reducing power curve shown in figure 13. The reported rate constant, k_r, was the slope of the best straight line through the experimental points, not including the initial point (0,100). Thus, k_r is the rate constant for the decrease in reducing power:

$$dR/dt = -k_r R \tag{8}$$

where

t = time, and

R = % of original reducing power.

Table 12.—Scope of data on glucose decomposition rates

Source	Sugar	Added acid concentration	Studies done at temperatures (°C) of								
			160	170	180	190	200	220	240	250	260
	%	%									
Saeman (1945)	5.0	0.4		*	*	*					
		.8		*	*	*					
		1.6		*	*	*					
Kirby (1948)	2.0	.4			*						
		.8			*						
		1.6	*	*	*	*					
		3.2			*						
	4.0	.4			*						
		.8			*						
		1.6	*	*	*	*					
		3.2			*						
	8.0	.4			*						
		.8			*						
		1.6	*	*	*	*					
		3.2			*						
McKibbins (1958)	5.0	1.0			*		*	*			
	10.0	.125							*	*	*
		.25						*	*	*	
		.50					*	*	*		
		1.0			*	*	*	*			
		2.0			*	*	*	*			
		4.0	*		*	*		*			
	20.0	1.0			*		*	*			

The presence of reversion material must be recognized when measurements of the rate of glucose degradation are interpreted. In such studies, glucose concentration is normally measured by a copper reduction procedure or, more recently, by a glucose oxidase enzymatic analysis. Neither method measures the total glucose (the sum of free and combined glucose) in solution. Using the reduction method, one can assay the reducing groups. Reducing end groups of oligomers are included. The glucose oxidase procedure measures only the monomeric or free glucose, and reversion material makes no contribution. When the rate of glucose decomposition is measured using either of these assay methods, an initial rapid decrease in glucose concentration is indicated. Such an initial rapid loss of free glucose was observed by Smith and coworkers (1982) when studying the short-time reactions of glucose in a flow reactor. They also observed a large amount of levoglucosan in the reversion material obtained from reaction at 220° C. This decrease must be ascribed to the formation of reversion products and not to the destruction (dehydration) of glucose. The rapid formation of reversion material, in an unequilibrated solution of glucose, simultaneously decreases both the reducing power and free glucose content of the solution. The magnitude of the effect can be quite large (fig. 13). Curves showing this phenomenon (fig. 13) were obtained from solution of the differential equations that describe the model in figure 11. The parameters used were $k_3/k_1 = k_5/k_1 = 35$, and k_1 was evaluated from McKibbins' (1958) correlation assuming a temperature of 220° C, acid concentration of 0.2N and initial glucose concentration of 1 M. The constants k_2 and k_4 were evaluated from k_3 and k_5 using the equilibrium constants of table 11. The hydrolysis rate, k_6, was assumed to equal k_3.

The rate of glucose degradation is almost equal to the rate of loss of either reducing power or free glucose after the reversion equilibrium has been established-i.e., after 10 seconds (fig. 13). The relationships between these rates (see App. D) are not linear; the free glucose decreases more slowly (1-20%) and the total glucose more rapidly (1-10%) than the reducing power, as the total concentration changes from 0 to 20%.

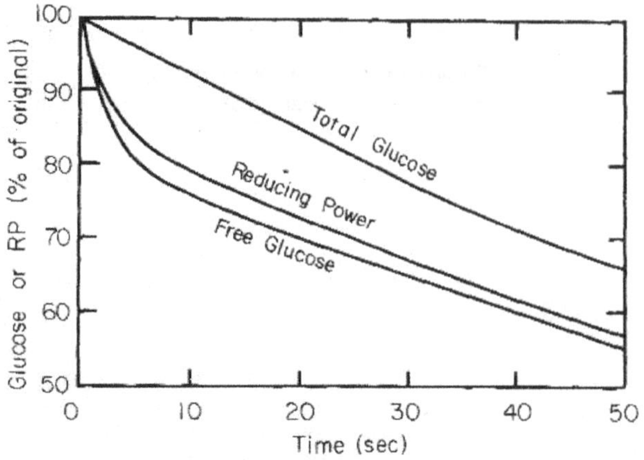

Figure 13.—Relationships between free glucose, reducing power (RP) and total glucose at 220°C, 0.2N H_2SO_4, 1M glucose. (ML84 5488)

It should be noted that McKibbins' correlation (1958; 1962) is based on the true normality of the solution. Solutions were prepared by adding the desired amount of glucose and acid (as 1M solution) to a volumetric flask and bringing to the desired volume with water. Thus, solutions of the same acid normality will have different acid-to-water ratios depending on the glucose concentration. In the studies by Kirby (1948) and Saeman (1945), solutions were prepared by dissolving the glucose in acid solutions of the desired strength. Using this procedure, the acid-to-water ratio is independent of glucose concentration. Kirby and Saeman found the glucose concentration to have no effect on their rate constants, whereas McKibbins found the rate constant to increase as glucose concentration increased. This apparent difference in the effect of the glucose concentration is due solely to the difference in defining acidity. During cellulose hydrolysis, the glucose content of the solution varies, and consequently, both the solution volume and acid normality will be time functions. This must be taken into account when calculating sugar yields.

Kinetics of Cellulose Hydrolysis

Of the three groups of data available (Kirby 1948; Saeman 1945; Saeman et al. 1950), the two more recent offer the best opportunity to interpret the effect of ash, acid concentration, and L/S on the rate of cellulose hydrolysis.

The substrate used for the experiments outlined in Kirby's thesis was a lignocellulose prepared, as he describes, by prehydrolyzing Douglas-fir wood. The effect of L/S on the rate of cellulose hydrolysis was found to be highly significant, especially at low acid concentration, This was properly ascribed to the ash content, which was found to be 0.38%. This same substrate was washed with HCl, lowering the ash content to 0.061%, and Saeman and coworkers (1950) ran a second set of experiments. Kirby correlated his original data set using three separate equations, one for each L/S, but the correlation was not satisfactory. The second set of data was not correlated. In all, 30 rate constants were reported. They contained a subgroup of 21 taken at 180° C (fig. 14). The large effect of L/S, especially at low added acid concentrations, is apparent. This data group was recorrelated to establish the effect of acidity on the cellulose hydrolysis rate in the following manner.

The rate constant was assumed to be related, at constant temperature, to the hydrogen ion concentration by the following equation:

$$k_c = Ax(CH)^n \qquad (9)$$

where

A and n = constants to be determined, and
CH = hydrogen ion concentration (mol/L).

For each of the two substrates, various neutralizing capacities 'were assumed. For an assumed neutralizing capacity, knowing the concentration of the added acid and L/S, CH can be calculated for each point using the procedure in Appendix A. The best straight-line fit of log k_c versus log (CH) was determined. Then the sum of squares of the differences between the line and the individual points was determined. The value of the neutralizing capacity resulting in the lowest value for the sum of squares was chosen as the neutralizing power of that substrate. The values determined in this manner were 0.110 eq/kg for the original substrate and 0.05 eq/kg for the acid washed substrates. The values of A and n obtained for each substrate were averaged to give the equation

$$k_c = 0.5107 \times (CH)^{1.218} \qquad (10)$$

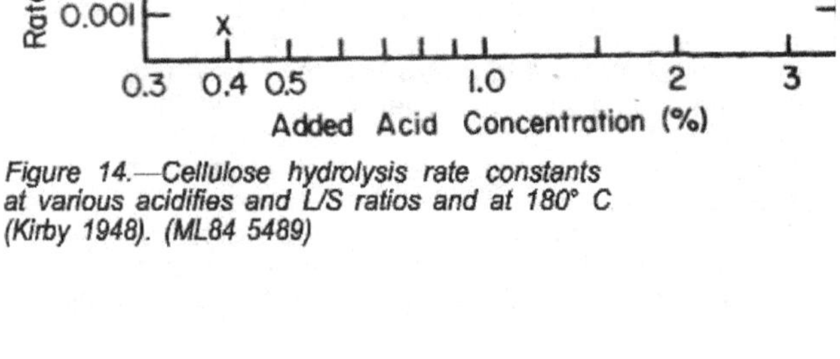

Figure 14.—Cellulose hydrolysis rate constants at various acidifies and L/S ratios and at 180° C (Kirby 1948). (ML84 5489)

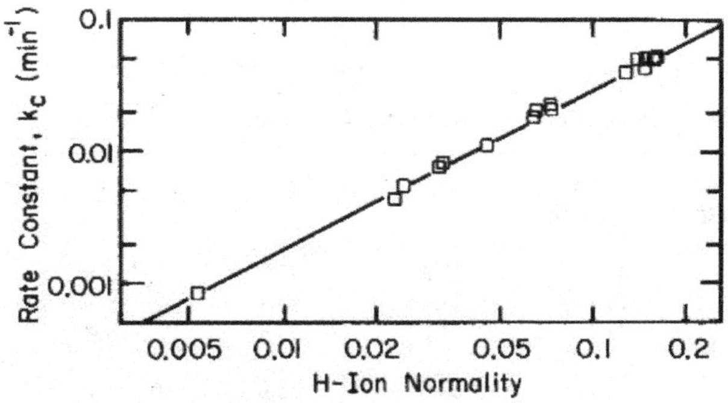

Figure 15.—Rate constants of figure 14 recorrelated using the true acidity and compared to eq. (10); $k_c = 0.5107 \times (CH)^{1.218}$. (ML84 5492)

A comparison of this equation with the same 21 data points of figure 14 is shown in figure 15. The correlation is quite good; the sum of squares of the deviations is 0.041, compared to 5.32 for Kirby's correlation. The maximum error for any of the points was 16%. There is, however, a disconcerting fact: The values of 0.110 and 0.05 eq/kg of substrate correspond to approximately 0.60 and 0.28% ash, whereas the measured values were 0.38 and 0.061%. The calculated values for neutralizing capacity stand in the ratio of approximately 2:1, while the measured values for ash have a ratio of 6:1. However, measurement of low ash quantities by dry ashing procedures is probably unreliable.

The value of the constant, A, is dependent on the temperature and would have the usual functional form

$$A = C \times \exp[-E_a/(RT)] \quad (11)$$

where

C = constant

E_a = activation energy (kJ/mol)

R = gas constant (8.314×10^{-3} kJ/mol °K)

T = temperature (K°)

Kirby (1948) reports a value of 188.7 and Saeman (1945) 179.6 kJ/mol as the activation energy for cellulose hydrolysis. Subsequent examination of glucose yields, described later, led to accepting Saeman's value, resulting in the following equation for the rate constant:

$$k_c = 2.80 \times 10^{20} \times (CH)^{1.218} \times \exp[-179.6/(RT)] \quad (12)$$

This is the rate constant for the resistant portion of the cellulose. However, all celluloses contain some material that is readily hydrolyzed being solubilized at about the same rate as hemicelluloses (Millett et al. 1954). Kirby's data indicate that, for his substrate, the readily removable cellulose was approximately 10% of the total cellulose. His prehydrolyzed Douglas-fir cellulose, which was dried, was probably similar to the dried material from prehydrolysis of southern red oak. The nature of this readily hydrolyzable material is not completely understood.

Glucose and Reversion Products from Cellulose

A model for glucose production from cellulose incorporates the previously described elements—cellulose hydrolysis, reversion, and glucose degradation (fig. 16). Its use is described in Appendix E.

Kirby (1948), Saeman et al. (1950), and Young (1949) reported a total of 59 data points for the maximum reducing power yields obtained at various reaction conditions with the Douglas-fir lignocellulose prepared by Kirby. These data are compared in table 13 with values calculated from the model by the procedure described in Appendix E using certain parameters:

initial cellulose content of the lignocellulose, CL = 57.9%,

resistant cellulose, CR = 90%,

neutralizing capacity of the residue, EQR = 0,

and those given in table 13. The fit is quite good except at the higher temperatures. However, those experimental points taken at short reaction times are probably in error because of the inadequacy of the technique as explained in the original publications. The model consistently predicts a much greater amount of cellulose remaining than that experimentally measured.

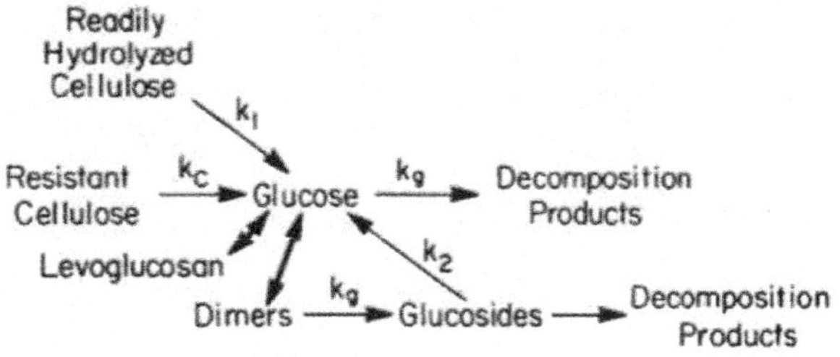

Figure 16.—Cellulose hydrolysis model. (ML84 5491)

Table 13.—Comparison of experimental[1] and calculated maximum yields from the hydrolysis of Douglas-fir lignocellulose

Temperature	Added acid concen-tration	L/S ratio	Time to maximum		Cellulose remaining		Reducing power yield		
			Experimental	Calculated	Experimental	Calculated	Experimental	Calculated	Difference
°C	%		------ min ------		------ % ------		------ % ------		
				HIGH-ASH LIGNOCELLULOSE (110 meq/kg)					
160	1.6	3	215.00	139.71	—	45.1	29.4	28.5	3.1
	1.6	6	185.00	122.30	—	44.1	31.4	29.8	5.1
	1.6	12	165.00	117.09	—	43.1	31.7	30.6	3.4
170	1.6	3	76.00	54.43	—	39.3	33.0	33.0	.1
	1.6	6	66.00	47.39	—	38.4	34.2	34.3	-.3
	1.6	12	59.00	45.14	—	37.5	34.7	35.1	-1.2
180	.8	3	80.00	60.46	—	42.1	30.0	31.4	-4.7
	1.6	3	26.00	21.15	—	35.0	36.3	37.5	-3.2
	3.2	3	11.20	8.39	—	31.9	38.0	40.7	-7.1
	.4	6	140.00	119.43	—	47.2	24.5	26.9	-9.7
	.8	6	60.00	44.68	—	38.4	34.8	34.5	.9
	1.6	6	22.00	18.79	—	33.4	37.2	38.7	-4.1
	3.2	6	11.10	7.95	—	31.2	39.3	41.7	-6.2
	.4	12	121.00	87.47	—	44.1	31.8	29.7	6.6
	.8	12	60.00	39.91	—	36.7	36.6	35.7	2.4
	1.6	12	22.00	17.43	—	33.4	38.2	39.5	-3.4
	3.2	12	10.70	7.84	—	30.5	40.0	42.4	-6.0
190	1.6	3	9.20	8.66	—	30.5	40.5	41.8	-3.3
	1.6	6	7.00	7.50	—	29.8	42.0	43.0	-2.4
	1.6	12	6.80	7.10	—	29.1	42.4	43.7	-3.0
				LOW-ASH LIGNOCELLULOSE (50 meq/kg)					
150	1.6	2	360.00	339.46	40.7	50.6	26.8	23.8	11.2
	1.6	6	360.00	307.02	38.0	49.5	28.0	25.6	8.5
	3.2	6	180.00	141.40	28.2	46.2	29.8	28.0	6.1
160	1.6	2	100.00	125.39	38.7	45.1	29.9	28.2	5.8
	1.6	2.50	160.00	120.05	—	45.1	30.0	28.6	4.5
	1.6	6	130.00	116.32	31.3	43.1	31.3	30.0	4.0
170	1.6	2	40.00	47.49	39.0	40.2	32.3	32.7	-1.3
	1.6	2.50	65.00	46.77	—	39.3	33.0	33.2	-.5
	.8	6	105.00	97.62	38.2	42.1	32.3	31.0	3.9
	1.6	6	40.00	43.66	39.2	38.4	34.4	34.5	-.4
	3.2	6	25.00	19.98	27.8	35.0	35.0	37.3	-6.6

(Page 1 of 2)

Table 13.—Comparison of expertmental[1] and calculated maximum yields from the hydrolysis of Douglas-fir lignocellulose—con.

Temperature	Added acid concen-tration	L/S ratio	Time to maximum		Cellulose remaining		Reducing power yield		
			Experimental	Calculated	Experimental	Calculated	Experimental	Calculated	Difference
°C	%		min		%		%		
				LOW-ASH LIGNOCELLULOSE (50 meq/kg)—con.					
180	1.6	2	22.00	18.98	—	35.0	35.5	37.3	- 5.0
	1.6	2	20.00	18.98	22.4	35.0	35.6	37.3	- 4.7
	.4	2.50	110.00	121.62	38.0	49.5	30.5	25.6	16.2
	1.6	2.50	20.00	18.61	—	34.2	38.5	37.7	2.0
	.4	3	148.00	109.59	—	47.2	25.8	26.9	- 4.1
	.8	3	56.00	43.28	—	38.4	34.3	34.0	1.0
	1.6	3	22.00	18.10	—	34.2	36.7	38.0	- 3.7
	1.6	6	15.00	17.31	32.6	33.4	38.3	39.0	-1.8
	.4	6	106.00	84.75	—	44.1	30.3	29.5	2.7
	.8	6	47.00	38.35	—	37.5	35.6	35.3	.8
	1.6	6	22.00	17.31	—	33.4	37.9	39.0	- 2.9
	3.2	6	10.80	7.81	—	30.5	39.5	41.9	-6.0
	1.6	12	22.00	16.76	—	33.4	38.7	39.6	- 2.4
190	1.6	2	6.00	7.78	27.5	30.5	38.8	41.7	- 7.6
	1.6	2.50	8.00	7.44	—	30.5	42.0	42.1	- .3
	.8	6	14.00	15.87	28.8	32.7	40.0	39.5	1.3
	1.6	6	6.00	7.05	26.7	29.1	41.6	43.3	- 4.0
	3.2	6	2.00	3.15	24.0	26.6	43.0	46.2	- 7.4
210	.4	2.50	11.00	9.95	29.0	35.0	37.7	36.5	3.3
	1.6	2.50	1.67	1.38	—	22.6	45.0	50.1	-11.2
230	.4	2.50	2.08	2.09	17.0	27.8	47.3	42.6	9.9
	1.6	2.50	.45	.28	—	17.1	50.5	56.0	- 10.9
240	1.6	2.50	.33	.13	—	14.9	51.5	58.0	- 12.5
250	.4	2.50	.63	.49	14.0	22.1	53.0	46.9	11.6
	1.6	2.50	.23	.06	—	13.0	48.5	59.2	- 22.0
260	.4	2.50	.45	.24	—	19.7	54.6	48.1	11.9
270	.4	2.50	.38	.13	—	17.1	53.7	48.6	9.4
280	.4	2.50	.30	.07	—	15.3	51.4	48.5	5.6

[1]Experimental values are from Kirby (1948), Saeman et al. (1950), and Young (1949).

(Page 2 of 2)

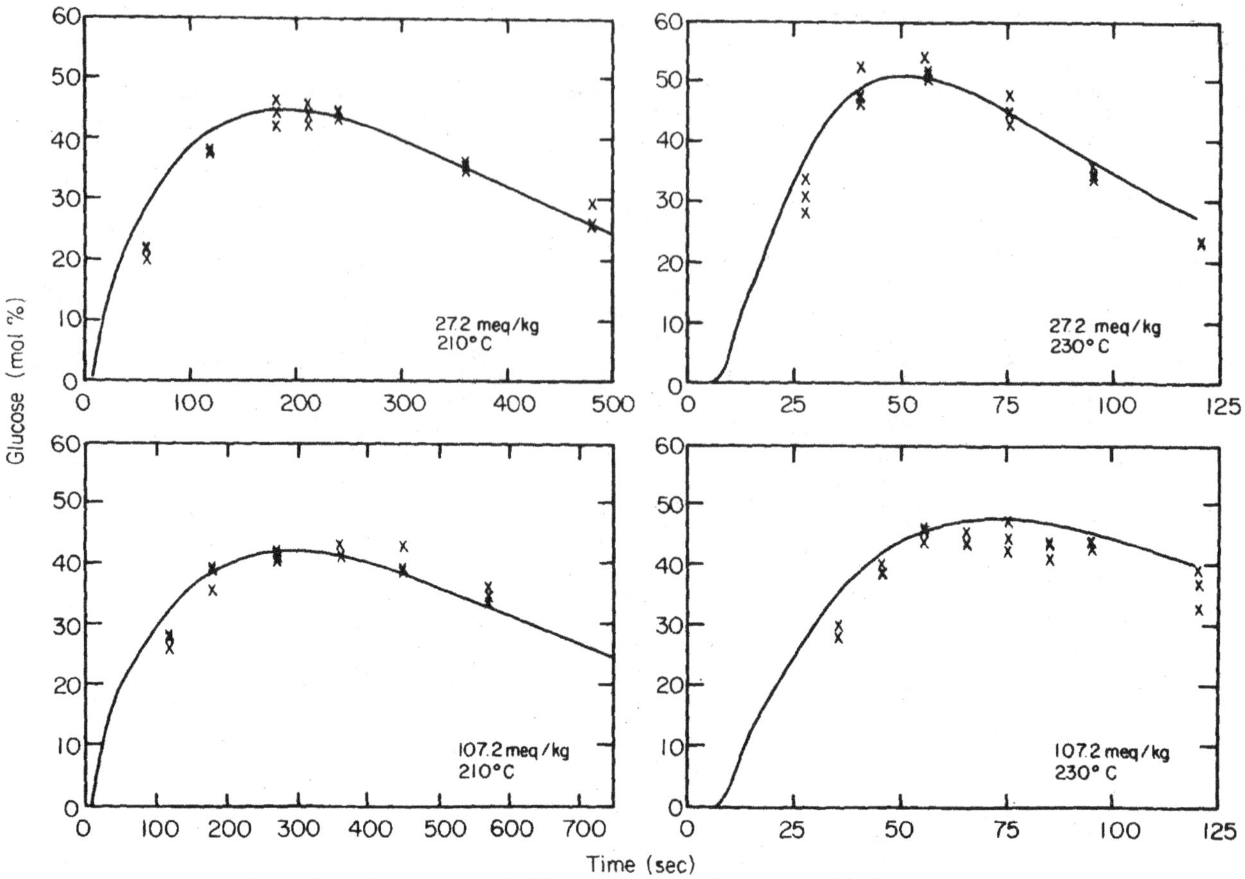

Figure 17.—Comparison of experimental glucose yields (% anhydroglucose based on original cellulose) from hydrolysis of prehydrolyzed southern red oak with glucose yields predicted by the model. Reaction conditions were 0.8% H_2SO_4 (added), L/S = 3. (ML84 5462)

We have obtained additional data using as substrate the residue from the prehydrolysis of southern red oak. These data (free glucose, reducing power and cellulose remaining) are shown in figures 17, 18, and 19, along with curves calculated from the model using the previously listed parameters.

The substrates were obtained from prehydrolysis run 19 (tables 6,7,8). The residue from run 19 was divided into two parts, one portion was washed with distilled water, the other with tap water. This resulted in two substrates, identical in composition (ash-free basis) but differing in ash content and consequently in neutralizing capacity. On an ash-free basis the material was analyzed:

lignin - 32.8%,

glucan - 58.1%,

xylan - 3.2%,

mannan - 0.3%, and

uronic anhydride - 0.6%.

The material washed with tap. water had an ash content of 0.42% and a neutralizing capacity (by ash titration) of 107.2 meq/kg OD material. Similar values for the distilled water washed substrate were 0.16% and 27.2 meq/kg. Both substrates were dried and fiberized. Reactions were carried out in glass ampoules (5-mm diameter) loaded with 0.2 g of the oven-dried fiberized substrate and 0.6 g of a 0.8% H_2SO_4 solution (see App. G). Analysis for the reducing power was by the Schaffer-Somogyi method and monomeric glucose determined by a newly developed liquid chromatographic procedure. The residual cellulose content was obtained from glucan analysis of the washed and dried solids remaining after hydrolysis.

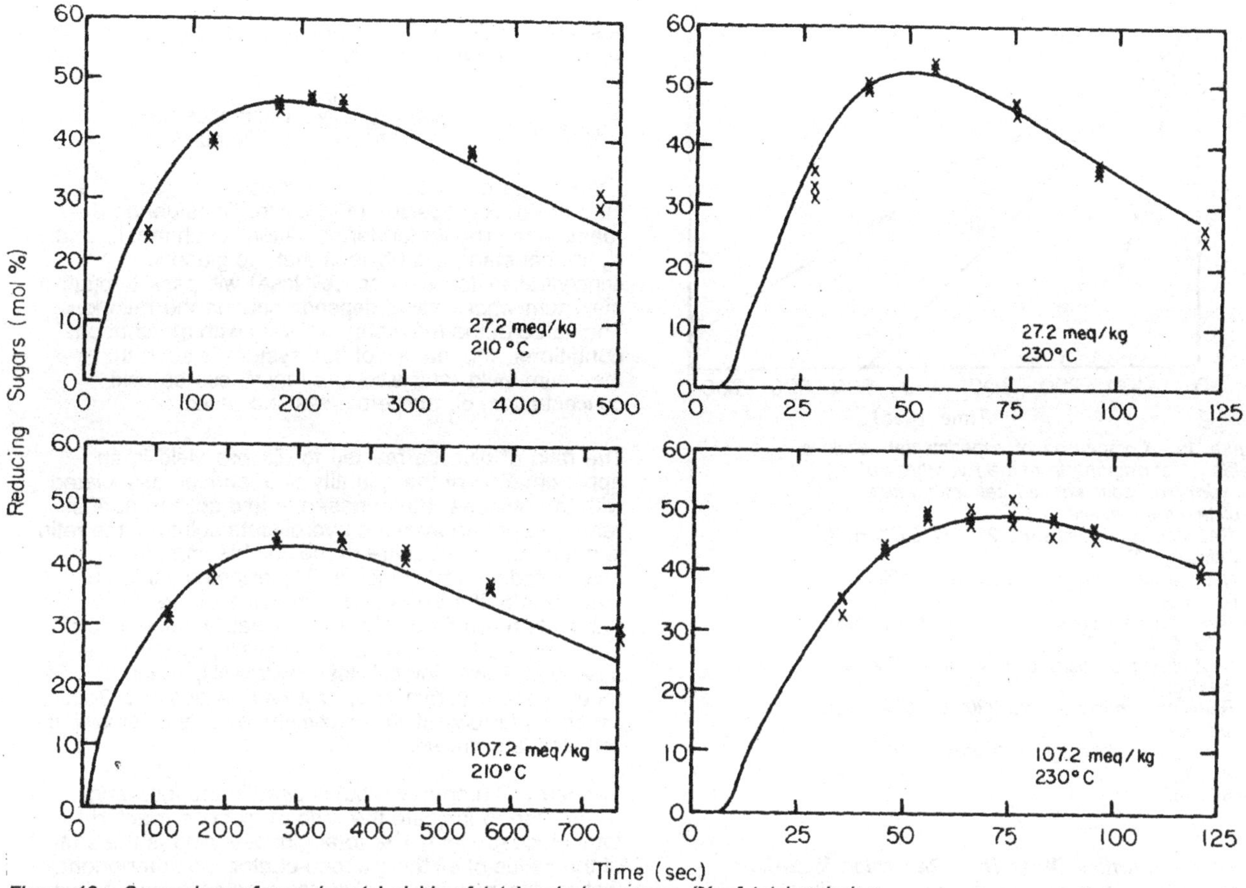

Figure 18.—Comparison of experimental yields of total reducing sugars (% of total reducing sugars in original prehydrolysis residue) from hydrolysis of prehydrolyzed southern red oak with yields predicted by the model. Reaction conditions were 0.8% H_2SO_4 (added), L/S = 3. (ML84 5463)

The curves in figures 17, 18, and 19 were obtained using the calculation procedure described in Appendix E with the following two modifications. Rather than assuming the nonresistant cellulose hydrolyzed instantaneously-i.e., k, (fig. 16) = ∞ —it was assumed to hydrolyze 50 times as fast as the resistant portion-i.e., k_1/k_c = 50. Also the reaction was not assumed to be isothermal but the temperature was assumed to rise in a manner previously established from the heat transfer characteristics of these small ampoules (see App. G).

Although no statistical analysis is presented, one will conclude from an examination of the figures that the calculated and measured values agree quite well. It may be concluded that the type of wood and the conditions used in the acidic prehydrolysis have little effect on the hydrolysis rate of the glucan. The model predicts satisfactory values for reducing power, free glucose, and cellulose yields as functions of time without modification of the parameters.

Table 14 shows the glucose yields calculated from the model for various levels of temperature, L/S, and concentration of applied acid. All yields are percent of theoretical-that is, percent of potential glucose available. These calculations assume isothermal conditions during the reaction and instantaneous hydrolysis of the nonresistant cellulose.

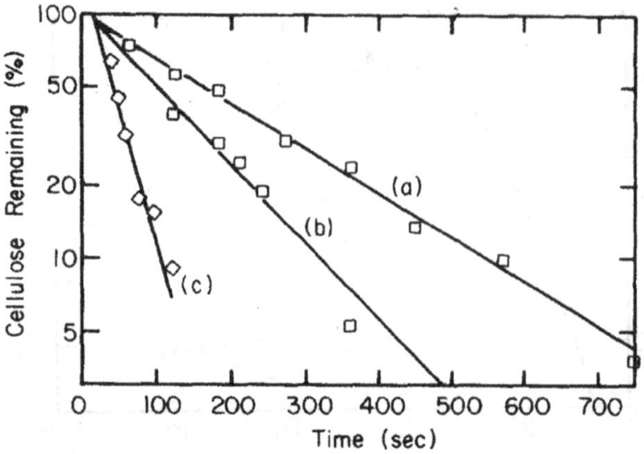

Figure 19.—Comparison of experimental cellulose yields (% of original) from the hydrolysis of prehydrolyzed southern red oak with yields predicted from model.
(a) Reaction conditions were 210° C, 0.8% H_2SO_4 (added), L/S = 3.
Neutralizing capacity of residue = 107.2 meq/kg.
(b) Reaction conditions were 210°C, 0.8% H_2SO_4 (added), L/S = 3.
Neutralizing capacity of residue = 27.2 meq/kg.
(c) Reaction conditions were 230° C, 0.8% H_2SO_4 (added), L/S = 3.
Neutralizing capacity of residue = 107.2 meq/kg.
(ML84 5480)

Organic Impurities Other Then Reversion Material
Impurity load from mass balance. —In addition to glucose and reversion products, the second-stage hydrolysate will contain a large quantity of impurities-over 20% of the glucose. Their presence significantly affects the design and economics of the process. To make any decisions regarding the second-stage design, it is necessary to know the nature and quantity of the various impurities. What follows is a first estimate of the variation with sugar yield of the amounts and quantities of the known impurities. The results can be used to make preliminary evaluations of possible process schemes and will also indicate what experimental work is needed.

Some revealing conclusions can be obtained by considering the mass balance. In the process we consider here, in contrast to percolation processes, the saccharification is done in a single-pass reaction (either batch or continuous) without removal of any of the constituents from the reaction zone. Since it is not economically feasible to recycle the solid because of its high lignin content, the reactor would be operated near maximum glucose yield.

A simple model for the hydrolysis of cellulose is a sequence of two reactions:

$$\text{Cellulose} \xrightarrow{k_c} \text{Glucose} \xrightarrow{k_g} \text{Decomposition Products}$$

There is no evidence that impurities arise from other than the decomposition of glucose. Considering the reaction to proceed under conditions such that k_c and k_g are constant, it is obvious that the glucose concentration (or yield on cellulose) will pass through a maximum whose value depends only on the ratio k_c/k_g. The value of the maximum will vary with the chosen conditions. The nature of this system is such that maximum yield (and also k_c/k_g) increases as acid concentration or temperature-is increased.

The ratio of glucose reacted to glucose yield is an approximation of the quantity of impurities associated with the glucose. Since gaseous and solid impurities readily separate from the hydrolysate solution, the ratio is not an exact measure of the soluble impurities associated with the glucose. Nevertheless, it is an indication of the trend and magnitude of the purification required. The ratio is easily calculated if k_c and k_g are known or, more precisely, by using the described model for cellulose hydrolysis, which requires more parameters, or it can be obtained from experimental data if the maximum yield and remaining cellulose are known.

The term IR (impurity ratio) is used in the following discussion to indicate the ratio of glucose reacted to total glucose yield. The total glucose yield is the sum of the yields of all the glucose-containing components—that is, the monomeric glucose, oligomers, and anhydro sugars. The numerator is the unaccounted glucose; it is obtained by subtracting the sum of the total glucose and the potential glucose of the unreacted cellulose from the potential glucose of the original cellulose charged. The value of IR at the time of maximum total glucose yield is of particular interest; this value is denoted by $(IR)_m$.

Both the maximum total glucose yield and $(IR)_m$ are dependent on the particular conditions of hydrolysis. Their relationship was calculated (fig. 20) by assuming the simple consecutive reactions model for the cellulose hydrolysis, using values for the rate constants previously presented. It is apparent that, at high maximum glucose yields, $(IR)_m$ is significantly less than at low yields. Increasing the maximum yield from 45 to 65% drops $(IR)_m$ from 0.51 to 0.33% or 35%. It is also apparent that even at the higher yields the reactants contain large quantities of materials other than glucose.

Table 14.—Maximum glucose yields from southern red oak lignocellulose calculated from model[1]

Temperature	Added acid concentration	L/S ratio	Glucose yields: Total	Glucose yields: Free	Glucose yields: Combined	Solution concentration	Time to maximum	Remaining cellulose	Residue yield
°C	%		%[2]			%	*min*	%	
200	.4	2	27.08	23.98	3.10	8.01	186.71	49.44	24.89
		3	32.95	29.52	3.44	6.60	32.01	42.56	30.44
		4	35.90	32.47	3.44	5.46	22.73	38.90	33.32
		5	37.59	34.24	3.35	4.61	19.06	37.75	35.00
	.8	2	42.21	36.24	5.97	11.95	11.17	34.51	38.26
		3	45.04	39.63	5.41	8.81	8.96	30.61	41.28
		4	45.62	40.79	4.84	6.84	8.01	29.71	42.12
		5	45.95	41.51	4.44	5.58	7.42	29.71	42.62
	1.6	2	50.57	42.76	7.80	13.99	3.70	27.15	45.52
		3	50.63	44.20	6.43	9.79	3.34	26.35	46.24
		4	50.65	45.02	5.63	7.53	3.23	25.57	46.63
		5	50.66	45.56	5.10	6.12	3.12	25.57	46.89
	3.2	2	54.53	45.80	8.73	14.92	1.44	24.09	48.94
		3	54.21	47.10	7.11	10.41	1.40	23.38	49.40
		4	54.06	47.87	6.19	8.00	1.36	23.38	49.69
		5	53.98	48.39	5.58	6.49	1.34	23.38	49.89
210	.4	2	30.93	27.10	3.84	9.05	83.49	45.19	28.05
		3	37.13	32.89	4.24	7.37	14.44	37.75	33.84
		4	40.18	35.92	4.27	6.07	10.10	34.51	36.78
		5	41.91	37.71	4.20	5.12	8.44	33.49	38.47
	.8	2	46.86	39.99	6.87	13.10	5.02	29.71	42.00
		3	49.67	43.35	6.32	9.63	3.87	27.15	44.97
		4	50.19	44.43	5.76	7.47	3.45	26.35	45.73
		5	50.48	45.11	5.37	6.10	3.27	25.57	46.19
	1.6	2	55.42	46.70	8.72	15.13	1.62	23.38	49.40
		3	55.31	47.96	7.35	10.60	1.46	22.69	49.94
		4	55.24	48.68	6.56	8.16	1.38	22.69	50.23
		5	55.20	49.15	6.05	6.63	1.36	22.02	50.43
	3.2	2	59.38	49.76	9.62	16.04	.63	20.74	52.80
		3	58.87	50.85	8.02	11.21	.60	20.12	53.06
		4	58.62	51.50	7.12	8.62	.59	20.12	53.24
		5	58.48	51.94	6.54	7.00	.58	20.12	53.37
220	.4	2	34.90	30.13	4.77	10.09	38.13	41.30	31.11
		3	41.33	36.07	5.26	8.14	6.44	34.51	37.02
		4	44.43	39.10	5.34	6.67	4.59	30.61	39.94
		5	46.18	40.88	5.30	5.61	3.82	29.71	41.62
	.8	2	51.42	43.44	7.99	14.19	2.25	26.35	45.41
		3	54.15	46.67	7.48	10.40	1.76	23.38	48.23
		4	54.58	47.65	6.93	8.07	1.56	22.69	48.90
		5	54.82	48.26	6.56	6.59	1.45	22.69	49.29
	1.6	2	60.07	50.21	9.86	16.19	.71	20.74	52.80
		3	59.77	51.25	8.52	11.36	.65	19.53	53.12
		4	59.60	51.84	7.76	8.75	.62	19.53	53.30
		5	59.49	52.22	7.27	7.11	.60	19.53	53.42
	3.2	2	63.98	53.24	10.74	17.07	.28	17.85	56.13
		3	63.26	54.08	9.19	11.95	.26	17.85	56.15
		4	62.91	54.59	8.33	9.19	.26	17.33	56.21
		5	62.70	54.93	7.78	7.47	.26	17.33	56.26

(Page 1 of 2)

Table 14.—Maximum glucose yields from southern red oak lignocellulose calculated from model*-con.

Temperature	Added acid concentration	L/S ratio	Glucose yields Total	Glucose yields Free	Glucose yields Combined	Solution concentration	Time to maximum	Remaining cellulose	Residue yield
°C	%		———— %[2]	————	————	%	min	——— %	———
230	.4	2	38.92	32.97	5.95	11.13	18.43	36.64	33.95
		3	45.50	38.93	6.56	8.89	3.03	30.61	39.86
		4	48.60	41.90	6.70	7.25	2.14	27.15	42.71
		5	50.33	43.62	6.70	6.08	1.77	26.35	44.34
	.8	2	55.85	46.46	9.38	15.23	1.03	23.38	48.34
		3	58.43	49.50	8.93	11.13	.80	20.74	50.97
		4	58.77	50.35	8.42	8.64	.71	20.12	51.51
		5	58.95	50.87	8.08	7.05	.67	19.53	51.83
	1.6	2	64.47	53.19	11.28	17.18	.33	17.85	55.61
		3	63.96	53.96	10.00	12.06	.30	17.33	55.70
		4	63.69	54.40	9.29	9.29	.28	16.81	55.76
		5	63.52	54.69	8.83	7.56	.27	16.81	55.80
	3.2	2	68.29	56.14	12.15	18.01	.13	15.37	58.82
		3	67.36	56.68	10.67	12.62	.12	15.37	58.59
		4	66.90	57.04	9.87	9.71	.12	14.92	58.52
		5	66.63	57.27	9.35	7.90	.12	14.92	58.48
240	.4	2	42.95	35.52	7.44	12.14	8.78	33.49	36.47
		3	49.58	41.38	8.20	9.61	1.46	27.15	42.26
		4	52.63	44.22	8.41	7.80	1.02	24.09	44.99
		5	54.32	45.85	8.46	6.53	.86	22.69	46.52
	.8	2	60.09	48.96	11.13	16.20	.49	20.74	50.72
		3	62.47	51.72	10.75	11.81	.38	17.85	53.08
		4	62.71	52.43	10.29	9.16	.34	17.33	53.49
		5	62.82	52.85	9.97	7.48	.31	17.33	53.73
	1.6	2	68.60	55.52	13.08	18.08	.16	15.37	57.74
		3	67.87	56.01	11.86	12.71	.14	14.92	57.58
		4	67.49	56.29	11.20	9.79	.13	14.47	57.52
		5	67.26	56.48	10.78	7.96	.13	14.47	57.48
	3.2	2	72.30	58.35	13.95	18.87	.06	13.23	60.79
		3	71.14	58.59	12.55	13.24	.06	13.23	60.31
		4	70.58	58.78	11.80	10.19	.06	12.84	60.11
		5	70.23	58.91	11.33	8.29	.06	12.84	59.99
250	.4	2	46.95	37.67	9.28	13.12	4.40	29.71	38.58
		3	53.53	43.31	10.22	10.30	.72	24.09	44.13
		4	56.50	45.99	10.51	8.33	.50	21.37	46.69
		5	58.12	47.50	10.62	6.96	.42	20.12	48.11
	.8	2	64.11	50.84	13.27	17.10	.24	17.85	52.44
		3	66.27	53.28	12.99	12.44	.18	15.84	54.50
		4	66.40	53.83	12.57	9.65	.16	15.37	54.78
		5	66.44	54.15	12.29	7.88	.15	14.92	54.93
	1.6	2	72.44	57.14	15.30	18.90	.08	13.23	59.13
		3	71.48	57.33	14.16	13.29	.07	12.84	58.73
		4	71.00	57.46	13.54	10.25	.06	12.46	58.54
		5	70.70	57.54	13.16	8.34	.06	12.46	56.43
	3.2	2	75.98	59.81	16.18	19.64	.03	11.74	61.97
		3	74.61	59.74	14.86	13.79	.03	11.39	61.26
		4	73.94	59.77	14.17	10.63	.03	11.05	60.93
		5	73.53	59.79	13.74	8.64	.03	11.05	60.74

[1]Lignocellulose is 57.9% cellulose and 80 meq/kg neutralizing capacity.

[2]Based on potential glucose available from the lignocellulose.

(Page 2 of 2)

During the hydrolysis of cellulose under fixed reaction conditions IR increases continuously. it is initially zero and rises toward infinity as the cellulose and glucose disappear. During reaction, the total glucose yield increases from zero initially to a maximum value and then continuously decreases. The relationship between total glucose yield and IR (up to the maximum glucose yield) is shown (fig. 21) for the particular conditions—230° C, 0.8% acid, and L/S = 3, employing a substrate of 57.9% cellulose and a neutralizing capacity of 80 meq/kg. The maximum yield of 58.5% total glucose is reached in about 50 seconds. IR rises continuously as the total glucose yield increases but increases very sharply as the maximum total glucose yield is approached. At the maximum yield of 58.5% glucose, the impurity ratio is 0.37; if the reaction were stopped at a 57.9% yield, a 1% reduction from the maximum, the IR would be 0.29. Taking a 2% reduction in sugar yield drops the ratio to 0.26, a 30% reduction from that at maximum glucose yield.

It is obvious, considering figure 20, that selection of conditions of temperature and acid that result in high yields also has the beneficial result of improving product quality. However, it is also clear from figure 21 that optimum operation will not be at the maximum yield time, but probably somewhat less.

Table 13 contains experimental data from Kirby (1948), Saeman and coworkers (1945; 1950), and Young (1949) that include the amount of cellulose remaining at maximum reducing power yield. For these data points, $(IR)_m$ can be obtained if correction is made for the reversion products. The ratio calculated from these experimental values was compared with that predicted from the previously described model. in each case, the experimental ratio was higher than that from the model. This is due largely to the greater amount of cellulose degradation observed over that predicted (table 13). On the average, the experimental ratio was 22% greater than that calculated.

Impurities from glucose decomposition.—The best information available to quantitatively estimate the composition of impurities resulting from hydrolysis is McKibbins' work on glucose decomposition (McKibbins 1958; 1962). included in this work are measurements of the hydroxymethylfurfural (HMF), levulinic acids (LA), total organic acids, and precipitated solids formed from reaction of 0.556M glucose at 220° C and 0.2N H_2SO_4 (figs. 22 and 23).

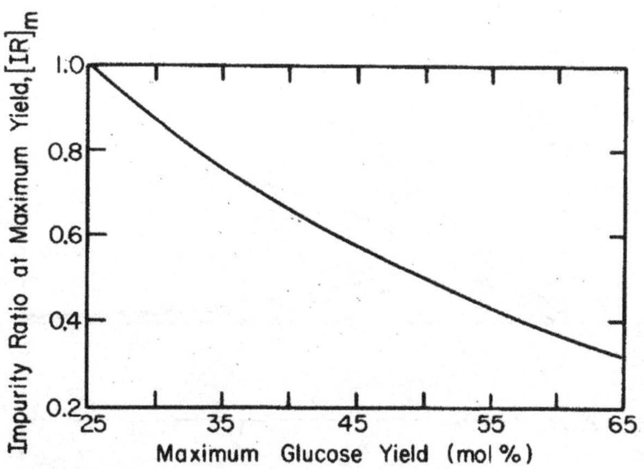

Figure 20.—Calculated variation of the impurity ratio (glucose reacted/glucose yield), $(IR)_m$, at maximum yield. (ML84 5481).

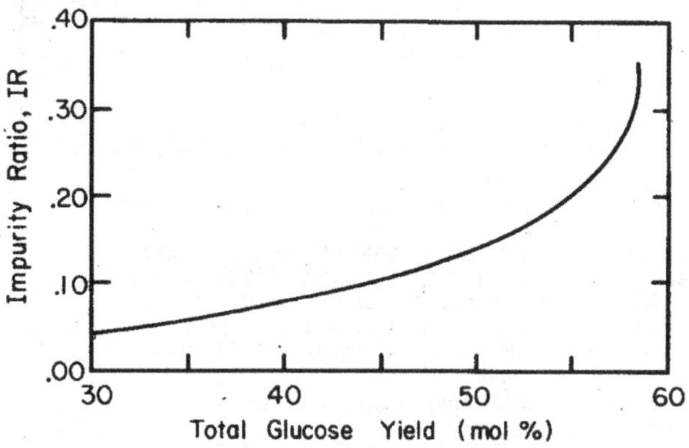

Figure 21.—Calculated change in the impurity ratio (glucose reacted/glucose yield), IR upon approaching maximum glucose yield. Conditions were 230° C, 0.8% H_2SO_4 (added), L/S = 3. (ML84 5483)

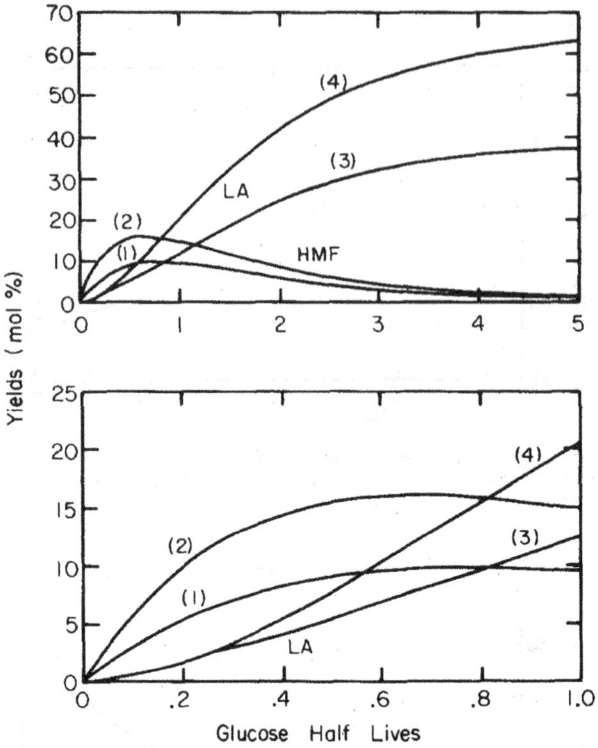

Figure 22.—Yields of hydroxymethylfurfural (HMF) and levulinic acid (LA) from glucose, 220°C, 0.2N H_2SO_4 0.556M glucose. Glucose half-lives is a measure of the extent of glucose decomposition.
Curve (1) is HMF yields from McKibbins' (1958) correlation of experimental values
Curve (2) is HMF yields assuming simple consecutive reactions
Curve (3) is LA yields from McKibbins' (1958) correlation of experimental values
Curve (4) is LA yields assuming formation from HMF present in the concentration of curve 1.
The lower half of this figure is a portion of the upper, on an expanded scale. (ML84 5464)

McKibbins measured the rate of both HMF and glucose degradation at the above conditions. If the mechanism of HMF formation is assumed to be simply a pair of consecutive first-order reactions-that is, formation and degradation-the resulting yield curve would be that shown in curve (2) (fig. 22). The actual yield follows curve (1), which lies below curve (2). This can be interpreted to mean that HMF reactions are somewhat similar to the furfural reactions where the loss is primarily incurred by second-order reactions of the product with precursors-that is,

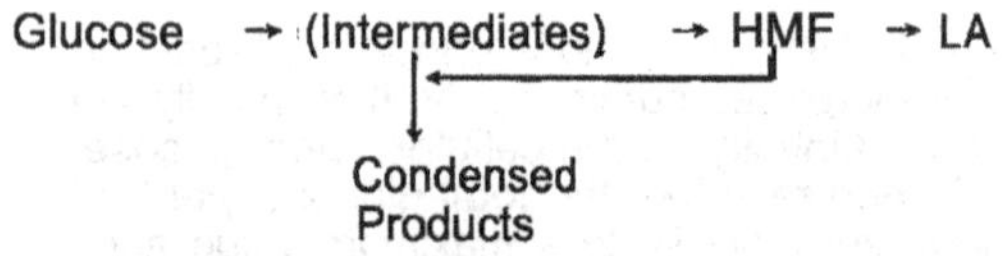

However, unlike furfural, the greatest loss of HMF is not through the formation of condensed products but, rather, due to its instability. For instance, at 0.65 glucose half-life (maximum HMF yield), the yield of HMF, if it were completely stable, is calculated to be 36%; curve (2) indicates ~16 % and curve (1), ~9.6%. Thus, 55% of the potential HMF either reacted with itself or formed levulinic acid, 18% polymerized with precursors, and 27% remains as HMF.

The appearance of LA is shown in curves (3) and (4) (fig. 22). Curve (4) indicates the quantity of LA that should be formed from the actual HMF (curve (1)) in solution reacting at the measured rate of HMF degradation. Curve (3), that actually obtained, shows much lower values. Although only vaguely indicated in figure 22 upper, it should be noted that the yield curves for LA reach a plateau and do not noticeably decrease, indicating that LA does not decompose or react with itself. The large loss of LA--that is, the difference between curves (3) and (4), can be explained by assuming that LA reacts extensively with its precursors or HMF reacts with itself. These reactions drop the yield of LA from a potential of ~65% (at 5.0 half-lives) to ~35%.

It is expected that the conditions of the second-stage hydrolysis will be in the vicinity of 230° C and 0.8% H_2SO_4. Under these conditions, maximum glucose yield is reached in about 1 glucose half-life. HMF is the predominant impurity early in the reaction, but the LA exceeds it at maximum glucose yield (fig. 22 lower).

McKibbins (1958) measured the rise of acidity as the glucose reaction proceeded (fig. 23 upper; curve (2)). Beyond the range of the plot, acid production decreases slowly, which is thought to be due to the degradation of formic acid; if this is true, it can be shown, using data collected by Dunaway (1950), that not more than 10% of the formic acid formed could be degraded over the range shown here. Thus, for our purpose, we can suppose the formic acid to be stable. Curve (1) of figure 23 upper is the sum of the acidity of LA and formic acid. The LA component is the actual measured quantity (fig. 22, curve (3)), whereas the formic component is the amount that would be generated along with the LA of curve (4), in figure 22. The rationale for doing this is the supposition that the LA yield is reduced, from curve (4) to curve (3) (fig. 22), by the interaction of LA with its precursors, but the formic does not undergo such reactions. By comparing curves (1) and (2) in figure 23 upper, it is apparent that making this assumption accounts for the total acidity. The experimental value would be expected to be slightly lower because of the instability of the formic acid.

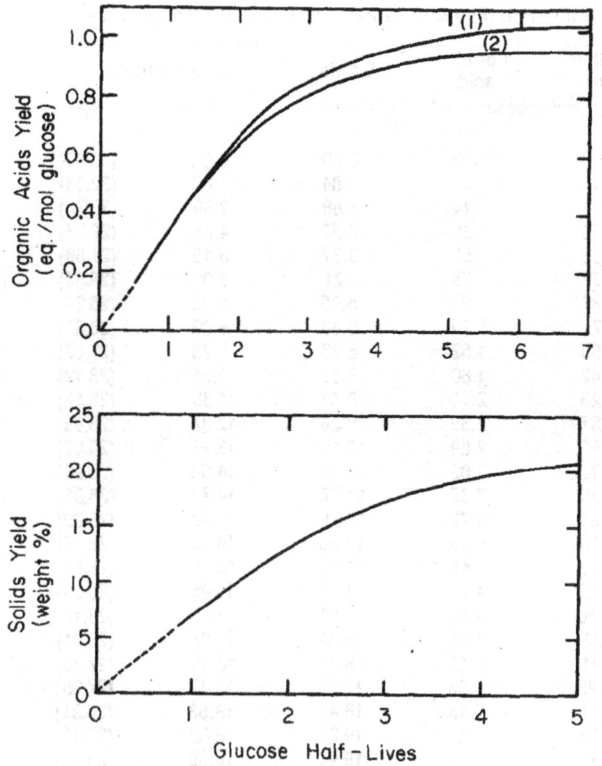

Figure 23.—Yields of organic acids (upper) and of solids (lower) from glucose, 220° C, 0.2N H_2SO_4, 0.556M glucose.
Upper: Curve (1) is formic acid generated from HMF present in the concentration of curve (1) of figure 22, plus the levulinic acid measured experimentally (curve (3) of fig. 22).
Curve (2) is total acids (McKibbins 1958).
Lower: Solids (McKibbins 1958).
(ML84 5467)

In the subsequent calculations, it is assumed that only two acids are present—levulinic and formic. The amount of formic acid was calculated by subtracting the equlvalents of levulinic acid formed from the total acid yield. If it is true that the total acidity is accounted for by the formic and levulinic acids, it would follow that HMF does not form significant quantities of dimers or higher oligomers-that is, it does not react with itself-and the major loss of LA must occur through reaction with precursors.

Solids formation accompanies glucose degradation (fig. 23 lower). As is the case with all the previous curves except for HMF, most of the experimental points are above 1 glucose half-life, whereas our range of interest is from 0.5 to 1.0 half-life. One might suppose there would be a significant induction period for solids formation. However, if it is assumed that much of the solid is generated from HMF reactions, the appearance of solids would occur very early in the reaction since the HMF maximum occurs at ~0.6 half-life. The assumption was made that the solids increased linearly up to the first experimental value (6.42 weight % at 0.861 half-life). Only one elemental analysis of the solids composition is available; it supports the empirical formula $C_6H_4O_2$.

With the information on the yields of the known products, a material balance on the carbon can be made (table 15). The unaccounted material is presented on two bases-glucose charged and glucose reacted. Almost a third of the reacted sugar is unaccounted for.

The assumption that levulinic and formic are the only acids present results in the maximum amount of unaccountable material. If acids containing more than one carbon are present, the additional carbon must come from the unaccounted. The unidentified products are probably low-molecular-weight furan polymers and levulinic acid condensation products.

Impurities in the cellulose hydrolysate.—An approximation of the hydrolysate composition can be based on the preceding information. However, its reliability is uncertain because we do not have a differential model for the degradation of glucose and the production of the various impurities. In the case of the glucose degradation studies and the information in table 15, the extent of glucose degradation and the time of reaction are related by the equation

$$C/C_0 = \exp(-kt) \qquad (13)$$

where

C = glucose concentration

C_0 = initial glucose concentration

t = time

As time proceeds, C decreases in an exponential manner and there is a particular relationship between the concentrations of the various reacting intermediates and products. This relationship will be different for the degradation of glucose generated from cellulose. Here the glucose concentration, initially zero, increases to a maximum and then decreases. Obviously, the concentrations and therefore the extent of reaction of, the various solution components will be different.

Table 15.—Distribution of carbon during glucose decomposition for reaction conditions of 220°C 0.2N H_2SO_4, 0.556M glucose[1]

Time (half-lives)	Free glucose	Reversion product	HMF	Levulinic acid	Formic acid	Solids	Unaccounted	
	% of carbon in original glucose							%[2]
0.00	86.51	13.49	0.00	0.00	0.00	0.00	0.00	(0.00)
.07	82.41	12.61	2.31	.19	.04	.84	1.60	(32.13)
.14	78.50	11.78	4.18	.72	.14	1.68	2.99	(30.76)
.22	74.77	11.03	5.69	1.53	.31	2.52	4.15	(29.23)
.29	71.21	10.32	6.88	2.56	.51	3.37	5.15	(27.88)
.36	67.82	9.66	7.81	3.75	.75	4.21	6.00	(26.64)
.43	64.59	9.05	8.52	4.87	.97	5.05	6.95	(26.37)
.50	61.50	8.47	9.04	5.7	1.24	5.89	8.09	(26.93)
.58	58.57	7.95	9.40	6.60	1.52	6.73	9.23	(27.57)
.65	55.76	7.46	9.64	7.42	1.80	7.57	10.34	(28.12)
.72	53.10	6.99	9.76	8.25	2.10	8.42	11.38	(28.52)
.79	50.55	6.56	9.80	9.07	2.39	9.26	12.37	(28.83)
.87	48.13	6.16	9.76	9.90	2.69	10.10	13.26	(29.02)
.94	45.82	5.79	9.67	10.72	3.03	10.94	14.03	(28.98)
1.01	43.62	5.44	9.52	11.55	3.37	11.77	14.73	(28.91)
1.08	41.52	5.12	9.34	12.37	3.72	12.51	15.42	(28.89)
1.15	39.52	4.81	9.13	13.20	4.06	13.26	16.02	(28.77)
1.23	37.62	4.54	8.89	14.02	4.41	14.01	16.51	(28.54)
1.30	35.81	4.27	8.63	14.85	4.74	14.75	16.95	(28.29)
1.37	34.09	4.02	8.37	15.70	4.89	15.50	17.43	(28.17)
1.44	32.44	3.78	8.09	16.55	5.03	16.24	17.87	(28.01)
1.51	30.88	3.57	7.81	17.41	5.18	16.99	18.16	(27.70)
1.59	29.39	3.37	7.53	18.26	5.33	17.73	18.39	(27.35)
1.66	27.97	3.18	7.25	19.11	5.48	18.48	18.53	(26.91)
1.73	26.62	2.99	6.96	19.96	5.63	19.22	18.62	(26.45)
1.80	25.33	2.83	6.69	20.81	5.78	19.97	18.59	(25.88)
1.88	24.11	2.67	6.42	21.66	5.93	20.71	18.50	(25.27)
1.95	22.94	2.52	6.15	22.52	6.08	21.46	18.34	(24.60)
2.02	21.83	2.38	5.89	23.32	6.25	22.13	18.21	(24.02)
2.09	20.77	2.25	5.64	23.80	6.52	22.81	18.41	(23.92)

[1]From McKibbins (1958).

[2]Values in parentheses are unaccounted material based on sugar reacted.

An approximation of the hydrolysate composition was made by assuming that the yield of each of the impurities from glucose degradation depended only on the glucose half-life (table 15). However, two different half-life times for the glucose degradation were used, resulting in two estimates of the hydrolysate composition (table 16). One value was based on actual time; knowing the half-life of glucose at the hydrolysis conditions, the half-life reaction time was assumed to be the actual time divided by this half-life. The second estimate was based on the extent of glucose reacted; for example, if 50% of the glucose liberated during the hydrolysis period had reacted, the half-life reaction time would be 1 half-life. It was also assumed, in this calculation, that the unaccounted material had a carbon content of 57%, the same as HMF.

The beneficial effect of raising the temperature, which increases yield, is apparent in table 16. The yield of impurities drops only from ~17 to ~11%, but because of the increased yields, the impurity ratio (which can be determined from the data in the table) falls from a high of 0.53 at 200° C to a low of 0.12 at 250° C. More than one-third of the soluble impurities consists of unknown compounds. However, as pointed out previously, the calculated quantity of unaccounted material is affected by the assumption that all the acidity is contributed by formic and levulinic acids. This assumption also affects the calculated total acids. Acids comprise more than one-third of the total impurities (table 16). This is a minimum value; it would be a larger ratio if the above assumption were not valid. The relationship between HMF and levulinic should also be noted. At the lower temperatures, where the Impurity ratio is high, levulinic predominates; but at 250° C, the HMF is substantially higher than the levulinic. In the probable range of commercial operation, the amounts of HMF and levulinic are approximately equal.

Table 16.—Yields and concentrations at maximum yield from lignocellulose hydrolysis[1]

Temperature	Added acid concentration	L/S ratio	Yields at maximum: Glucose[2] Total	Glucose[2] Free[3]	Glucose[2] Combined	Impurities[4]	Concentrations: Glucose[2] Total	Glucose[2] Free	Glucose[2] Combined	HMF[4]	Levulinic acid[4]	Formic acid[4]	Unaccounted[4]	Cellulose content of residue[4]
°C	%		%											
200	0.4	2	30.02	26.59	3.43	15.77	8.00	7.08	0.91	0.57	1.44	0.85	1.34	39.16
						15.96				1.01	1.18	.65	1.41	39.34
		3	36.55	32.74	3.81	16.62	6.59	5.90	.69	.47	.96	.60	.97	35.66
						18.66				.78	.81	.42	.99	35.82
		4	39.83	36.02	3.81	17.16	5.45	4.93	.52	.39	.73	.46	.76	33.61
						17.14				.63	.62	.32	.77	33.76
		5	41.71	38.00	3.72	16.79	4.61	4.20	.41	.33	.57	.35	.61	32.99
						16.78				.51	.49	.25	.61	33.13
	.8	2	46.84	40.22	6.62	15.84	11.94	10.25	1.69	.82	1.18	.70	1.34	31.13
						15.86				1.23	1.03	.49	1.30	31.25
		3	49.97	43.97	6.00	16.56	8.80	7.74	1.06	.61	.84	.49	.97	28.58
						16.59				.90	.74	.35	.93	28.70
		4	50.62	45.25	5.36	16.78	6.83	6.10	.72	.48	.65	.38	.75	27.96
						16.81				.70	.57	.27	.73	28.08
		5	50.98	46.06	4.92	16.55	5.57	5.04	.54	.39	.52	.30	.60	27.98
						16.59				.56	.46	.22	.58	28.10
	1.6	2	56.10	47.45	8.65	15.16	13.97	11.82	2.15	.91	1.04	.57	1.26	26.32
						15.19				1.27	.93	.41	1.18	26.41
		3	56.17	49.05	7.13	15.66	9.78	8.54	1.24	.65	.75	.41	.91	25.71
						15.69				.91	.67	.30	.85	25.80
		4	56.19	49.95	6.24	15.65	7.52	6.69	.83	.51	.58	.31	.70	25.71
						15.68				.70	.52	.23	.66	25.80
		5	56.21	50.55	5.66	16.16	6.11	5.50	.61	.42	.49	.27	.58	25.11
						16.20				.58	.44	.20	.55	25.21
210	.4	2	34.30	30.05	4.25	16.16	9.03	7.91	1.12	.64	1.39	.85	1.38	37.06
						16.24				1.08	1.16	.61	1.42	37.23
		3	41.19	36.49	4.70	17.11	7.36	6.52	.84	.52	.95	.59	1.00	32.95
						17.10				.83	.81	.41	1.00	33.10
		4	44.59	39.86	4.73	17.23	6.06	5.42	.64	.43	.71	.44	.77	31.01
						17.23				.66	.61	.30	.76	31.15
		5	46.51	41.85	4.66	16.74	5.11	4.60	.51	.36	.55	.33	.61	30.42
						16.75				.54	.47	.23	.59	30.55
	.8	2	52.00	44.38	7.62	15.93	13.08	11.17	1.92	.87	1.15	.66	1.33	28.02
						15.97				1.28	1.00	.46	1.27	28.14
		3	55.11	48.10	7.01	15.76	9.61	8.39	1.22	.64	.77	.42	.91	26.27
						15.79				.90	.68	.31	.86	26.37
		4	55.69	49.30	6.38	15.95	7.46	6.60	0.85	0.50	0.59	0.33	0.71	25.69
						15.99				.70	.53	.24	.67	25.79
		5	56.01	50.05	5.95	16.28	6.09	5.44	.65	.42	.49	.27	.59	25.10
						16.32				.58	.44	.20	.56	25.20
	1.6	2	61.48	51.82	9.67	14.45	15.11	12.73	2.38	.92	.95	.50	1.17	23.57
						14.45				1.26	.85	.35	1.09	23.65
		3	61.37	53.22	8.15	14.98	10.59	9.18	1.41	.67	.69	.36	.86	23.01
						14.99				.90	.62	.26	.79	23.09
		4	61.29	54.01	7.28	15.03	8.15	7.18	.97	.52	.54	.28	.66	23.01
						15.04				.70	.48	.20	.61	23.08
		5	61.24	54.53	6.71	15.51	6.62	5.90	.73	.43	.45	.24	.56	22.46
						15.53				.58	.41	.17	.52	22.53

(Page 1 of 3)

Table 16.—Yields and concentrations at maximum yield from lignocellulose hydrolysis'-con

Temperature	Added acid concentration	L/S ratio	Yields at maximum: Gucose[2] Total	Free[3]	Combined	Impurities[4]	Concentrations: Glucose[2] Total	Free	Combined	HMF[4]	Levulink acid[4]	Formic acid[4]	Unaccounted[4]	Cellulose content of resdue[4]
°C	%		%											
220	.4	2	38.70	33.42	5.28	16.22	10.07	8.70	1.38	.71	1.32	.83	1.37	35.03
						16.21				1.14	1.12	.57	1.39	35.18
		3	45.85	40.02	5.83	16.45	8.13	7.10	1.03	.56	.87	.52	.96	31.08
						16.46				.86	.75	.36	.94	31.21
		4	49.31	43.39	5.92	16.98	6.66	5.86	.80	.46	.67	.40	.76	28.55
						17.00				.69	.59	.28	.74	28.68
		5	51.25	45.37	5.88	16.38	5.60	4.96	.64	.39	.51	.30	.59	27.99
						16.42				.56	.45	.21	.57	28.11
	.8	2	57.07	48.21	8.86	15.11	14.18	11.98	2.20	.90	1.04	.56	1.25	25.74
						15.14				1.27	.92	.40	1.17	25.84
		3	60.08	51.79	8.29	15.30	10.39	8.96	1.43	.67	.72	.38	.88	23.52
						15.31				.91	.65	.28	.82	23.60
		4	60.57	52.88	7.69	15.47	8.06	7.04	1.02	.52	.56	.29	.68	22.98
						15.49				.71	.50	.21	.64	23.06
		5	60.83	53.55	7.28	15.31	6.58	5.79	.79	.43	.45	.23	.55	22.99
						15.32				.57	.40	.17	.51	23.07
	1.6	2	66.65	55.72	10.93	13.10	16.17	13.52	2.65	.91	.83	.41	1.03	21.57
						13.06				1.21	.72	.28	.96	21.63
		3	66.32	56.87	9.45	13.71	11.35	9.73	1.62	.67	.61	.30	.76	21.04
						13.68				.88	.54	.21	.71	21.09
		4	66.14	57.53	8.61	14.23	8.74	7.60	1.14	.53	.49	.24	.61	20.52
						14.20				.69	.44	.18	.57	20.58
		5	66.02	57.95	8.07	14.30	7.10	6.23	.87	.44	.40	.20	.50	20.52
						14.28				.56	.36	.14	.47	20.57
230	.4	2	43.17	36.57	6.59	16.66	11.11	9.41	1.70	.77	1.31	.81	1.40	32.34
						16.65				1.22	1.12	.56	1.40	32.49
		3	50.48	43.20	7.28	16.25	8.88	7.60	1.28	.60	.83	.48	.95	28.60
						16.28				.89	.72	.34	.91	28.72
		4	53.93	46.50	7.43	15.92	7.24	6.24	1.00	.49	.60	.34	.71	26.84
						15.96				.69	.53	.24	.67	26.95
		5	55.86	48.42	7.44	15.85	6.08	5.27	.81	.41	.48	.26	.57	25.70
						15.88				.57	.43	.19	.54	25.79
	.8	2	61.98	51.57	10.41	14.14	15.21	12.66	2.56	.91	.93	.49	1.15	23.59
						14.14				1.25	.83	.34	1.06	23.67
		3	64.84	54.93	9.91	14.20	11.12	9.42	1.70	.67	.64	.33	.80	21.51
						14.18				.89	.57	.23	.74	21.57
		4	65.22	55.88	9.34	14.38	8.63	7.39	1.24	.53	.50	.25	.62	21.00
						14.36				.69	.45	.18	.58	21.06
		5	65.42	56.45	8.96	14.67	7.04	6.08	.96	.44	.41	.21	.52	20.50
						14.65				.57	.37	.15	.48	20.55
	1.6	2	71.54	59.03	12.52	12.07	17.16	14.16	3.00	.89	.73	.35	.93	19.20
						12.01				1.16	.50	.24	.89	19.24
		3	70.98	59.88	11.10	12.77	12.05	10.17	1.88	.67	.55	.26	.69	18.71
						12.72				.85	.47	.18	.66	18.75
		4	70.68	60.38	10.31	13.30	9.28	7.93	1.35	.53	.44	.21	.56	18.24
						13.26				.67	.38	.15	.53	18.28
		5	70.49	60.69	9.80	13.41	7.55	6.50	1.05	.44	.36	.17	.46	18.23
						13.37				.55	.32	.13	.44	18.27
240	.4	2	47.64	39.40	8.24	16.05	12.12	10.02	2.10	.81	1.20	.72	1.35	30.47
						16.06				1.24	1.04	.49	1.31	30.60
		3	55.01	45.92	9.10	15.82	9.60	8.01	1.59	.64	.77	.43	.92	26.27
						15.85				.91	.69	.31	.87	26.37

(Page 2 of 3)

Table 16.—Yields and concentrations at maximum yield from lignocellulose hydrolysis[1]–con

Temperature	Added acid concentration	L/S ratio	Yields at maximum: Glucose[2] Total	Glucose[2] Free[3]	Glucose[2] Combined	Impurities[4]	Concentrations: Glucose[2] Total	Glucose[2] Free	Glucose[2] Combined	HMF[4]	Levulinic acid[4]	Formic acid[4]	Unaccounted[4]	Cellulose content of residue[4]
°C	%		%											
		4	58.41	49.08	9.33	15.83	7.80	6.55	1.25	.52	.58	.31	.70	24.03
						15.85				.71	.52	.23	.66	24.12
		5	60.29	50.90	9.39	15.64	6.53	5.51	1.02	.43	.46	.24	.56	22.97
						15.66				.58	.41	.18	.52	23.05
	.8	2	66.69	54.34	12.35	13.07	16.18	13.19	3.00	.90	.83	.41	1.03	21.57
						13.04				1.21	.71	.28	.96	21.63
		3	69.33	57.40	11.93	13.42	11.80	9.77	2.03	.67	.59	.29	.74	19.14
						13.38				.87	.51	.20	.69	19.18
		4	69.60	58.19	11.41	13.61	9.15	7.65	1.50	.53	.46	.22	.58	18.67
						13.57				.68	.40	.16	.54	18.72
		5	69.72	58.66	11.07	13.54	7.47	6.28	1.19	.44	.37	.18	.47	18.68
						13.50				.55	.32	.13	.44	18.72
	1.6	2	76.13	61.62	14.51	10.95	18.06	14.62	3.44	.85	.64	.29	.82	17.03
						10.86				1.09	.48	.19	.82	17.07
		3	75.32	62.16	13.16	11.75	12.69	10.47	2.22	.65	.49	.22	.62	16.59
						11.68				.81	.39	.15	.62	16.62
		4	74.91	62.48	12.43	12.30	9.78	8.16	1.62	.52	.40	.18	.50	16.15
						12.24				.65	.33	.13	.49	16.19
		5	74.65	62.68	11.96	12.46	7.96	6.68	1.28	.43	.33	.15	.42	16.15
						12.40				.53	.27	.11	.41	16.18
250	.4	2	52.08	41.79	10.29	15.88	13.10	10.51	2.59	.86	1.15	.66	1.33	28.02
						15.92				1.28	1.00	.46	1.27	28.14
		3	59.40	48.07	11.34	15.23	10.29	8.32	1.96	.66	.72	.38	.88	24.07
						15.25				.91	.64	.28	.82	24.15
		4	62.71	51.04	11.67	15.06	8.32	6.77	1.55	.52	.54	.28	.66	21.97
						15.06				.70	.48	.20	.61	22.04
		5	64.51	52.73	11.79	14.81	6.95	5.68	1.27	.43	.42	.22	.52	20.98
						14.80				.57	.38	.16	.49	21.04
	.8	2	71.16	56.43	14.73	12.30	17.08	13.55	3.54	.89	.75	.36	.95	19.19
						12.25				1.17	.62	.25	.90	19.23
		3	73.55	59.14	14.41	12.21	12.43	9.99	2.44	.66	.52	.24	.65	17.41
						12.15				.83	.42	.17	.63	17.44
		4	73.69	59.75	13.95	12.44	9.64	7.81	1.82	.52	.41	.19	.51	16.98
						12.38				.65	.34	.13	.50	17.01
		5	73.74	60.10	13.64	12.72	7.87	6.41	1.46	.44	.34	.15	.43	16.55
						12.67				.54	.28	.11	.41	16.58
	1.6	2	80.40	63.42	16.97	9.79	18.88	14.89	3.99	.79	.56	.24	.71	15.06
						9.68				1.01	.37	.15	.75	15.09
		3	79.34	63.63	15.71	10.70	13.28	10.65	2.63	.62	.44	.19	.55	14.66
						10.60				.77	.31	.12	.57	14.68
		4	78.80	63.78	15.03	11.02	10.24	8.29	1.95	.50	.35	.15	.44	14.65
						10.93				.61	.26	.10	.45	14.67
		5	78.47	63.87	14.60	11.48	8.33	6.78	1.55	.42	.30	.13	.38	14.26
						11.40				.51	.23	.09	.38	14.28

[1]Lignocellulose is 57.9% cellulose and 80 meq/kg neutralizing capacity.

[2]Glucose and combined glucoseare reported as glucose.

[3]Based on cellulose charged.

[4]The first value in each set is derived assuming half-life based on time; the second value is derived assuming half-life based on extent of degradation.

(Page 3 of 3)

The conditions that are probably near the economic optimum are 230° C, 0.8% H_2SO_4, and L/S = 3. The impurity ratio at maximum yield for these conditions is about 0.2 (impurities/total glucose; table 18). As described earlier, if the reactor is operated at somewhat less than maximum yield, this ratio will be decreased. From the relationship between yields of sugars and impurities as a function of time for the above conditions (table 17), we can calculate that the impurity ratio can be decreased by 18% by taking a 1% loss in yield, and by 35% by taking a 5% loss in yield. Also, the type of impurities shifts; HMF becomes dominant and there is proportionately less acid as the reaction time is shortened. Table 17 indicates that the probable impurity ratio will be 0.15-0.20. This, of course, is only the impurity load resulting from the glucose decomposition. There will be, in addition to this, considerable furfural, acetic acid, and some lignin-derived solubles. These will probably increase the total impurity load by an additional 20-30%.

The two estimates for the yields of HMF, LA, and formic acid, as given in table 17, are plotted in figure 24 along with some experimental yield data for these compounds. Conditions of hydrolysis were 230° C, 0.8% H_2SO_4, and L/S = 3. The estimates appear to give upper and lower bounds for the actual yields of HMF, but predicted yields for both LA and formic acid differ significantly from measured values. However, the experimental values for these acids may be suspect since they were measured with an analytical method (HPLC) still undergoing development. The hydrolysates from the prehydrolysis residue, from which these data were obtained, contain large amounts of acetic acid, which interfere with the procedure. The experimental data do confirm the rapid rise in the amounts of LA and formic acid, and they also support the supposition that formic acid is obtained in more than the equimolar amount of LA.

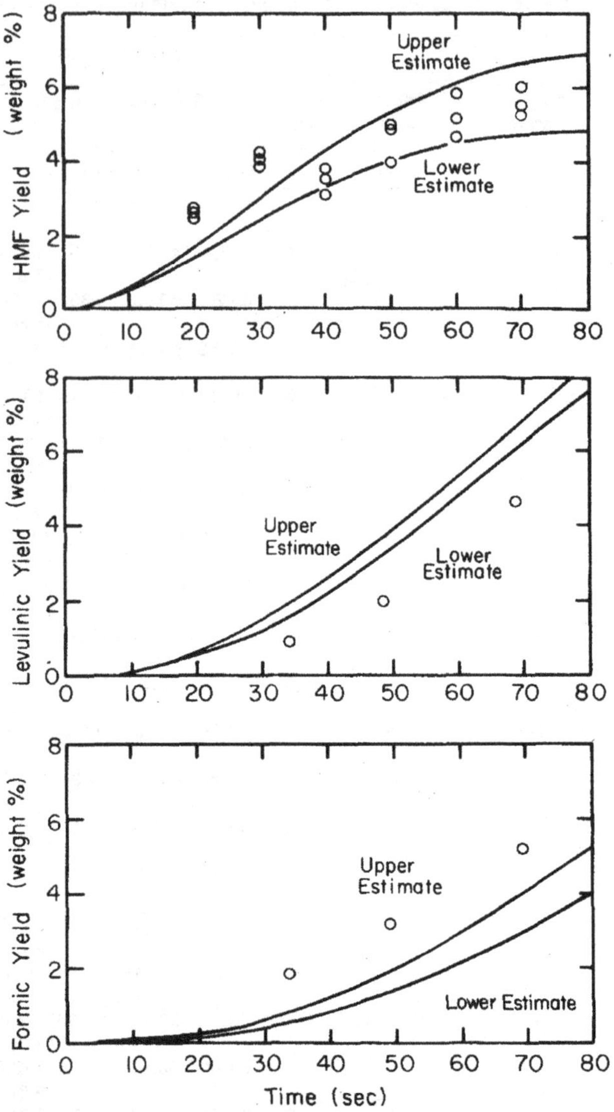

Figure 24.—Comparison of experimental and calculated yields of hydroxymethylfurfural. levulinic acid and formic acid at 230° C, 0.8% H_2SO_4 (added), L/S = 3. (ML84 5468)

Table 17.—Products from lignocellulose hydrolysis at reaction conditions of 230° C, 0.8% H_2SO_4, L/S = 3[1]

Time	Cellulose remaining	Free glucose	Disac-charides	Levo-glucosan	HMF	Levulinic acid	Formic acid	Solids	Unacounted
sec	*—% based on cellulose charged—*								
				HALF-LIFE BASED ON TIME					
1.0	87.3	12.3	0.17	1.34	0.02	0.00	0.00	0.01	0.02
7.9	70.8	27.1	.79	2.95	.38	.07	.03	.13	.27
14.7	57.4	37.7	1.50	4.11	.96	.30	.12	.37	.70
21.6	46.6	45.2	2.13	4.92	1.63	.79	.31	.72	1.24
28.5	37.8	50.1	2.60	5.45	2.31	1.47	.59	1.18	1.90
35.4	30.6	53.1	2.90	5.78	2.95	2.19	.97	1.73	2.78
42.3	24.8	54.6	3.06	5.94	3.52	2.98	1.43	2.36	3.76
48.2	20.7	54.9	3.10	5.98	3.91	3.74	1.90	2.96	4.65
55.0	16.8	54.5	3.06	5.94	4.30	4.67	2.49	3.70	5.69
61.9	13.6	53.5	2.94	5.82	4.57	5.66	3.18	4.48	6.68
68.8	11.1	51.9	2.78	5.65	4.74	6.71	3.96	5.28	7.63
75.7	9.0	50.0	2.59	5.45	4.83	7.78	4.76	6.06	8.52
82.6	7.3	47.9	2.38	5.22	4.88	8.87	5.57	6.86	9.31
89.4	5.9	45.7	2.17	4.97	4.84	10.02	6.18	7.67	10.07
96.3	4.8	43.4	1.97	4.72	4.74	11.19	6.68	8.50	10.75
				HALF-LIFE BASED ON GLUCOSE DEGRADATION					
1.0	87.3	12.3	.17	1.34	.02	.00	.00	.01	.02
7.9	70.8	27.1	.79	2.95	.40	.04	.01	.13	.28
14.7	57.4	37.7	1.50	4.11	1.06	.17	.07	.35	.75
21.6	46.6	45.2	2.13	4.92	1.90	.45	.18	.68	1.37
28.5	37.8	50.1	2.60	5.45	2.84	.91	.36	1.10	2.08
35.4	30.6	53.1	2.90	5.78	3.74	1.62	.64	1.60	2.81
42.3	24.8	54.6	3.06	5.94	4.57	2.53	1.00	2.16	3.56
48.2	20.7	54.9	3.10	5.98	5.19	3.34	1.34	2.68	4.31
55.0	16.8	54.5	3.06	5.94	5.80	4.21	1.83	3.33	5.33
61.9	13.6	53.5	2.94	5.82	6.30	5.07	2.37	4.01	6.41
68.8	11.1	51.9	2.78	5.65	6.62	5.98	2.96	4.73	7.49
75.7	9.0	50.0	2.59	5.45	6.82	6.91	3.60	5.47	8.54
82.6	7.3	47.9	2.38	5.22	6.97	7.85	4.23	6.21	9.51
89.4	5.9	45.7	2.17	4.97	6.96	8.83	5.02	6.99	10.35
96.3	4.8	43.4	1.97	4.72	6.87	9.83	5.82	7.73	11.15

[1]Lignocellulose is 57.9% cellulose and 80 meq/kg neutralizing capacity.

A number of important factors are relevant to the development of a complete process design. The final lesign depends on the choice of products-in addition to ethanol-to be marketed. This in turn depends on the impact of the process on the economy (market) and the environment,, as well as the availability of technology and the efficiency and economics of the process. For this two-stage process, consideration of such factors leads logically to the process design shown in figures 25 and 26.

First Stage

Hydrolyzer Type

If direct steam heating is used for the prehydrolysis, the movement of liquids during the heatup period leads to the selection of a continuous, tubular-type apparatus for the first-stage hydrolyzer. As previously described, during heatup some of the solution contained in the saturated chips moves from the interior of the solid to the surface where it mixes with condensate. In a thin bed, such as used in our experimental studies, the chips drain readily bringing the L/S of the heated mass to below its original value. Taking as an example run 19 (table 7), 100 kg of oak chips charged with an L/S of 1.27 would require approximately 50 kg of steam to raise them from ambient temperature to 170° C. The resulting condensate would raise the L/S to 1.77 if all the liquid remained in contact with the chips. However, in the thin bed the ratio drops to 1.05. Since this liquid movement occurs prior to any appreciable hydrolysis, if the excess is removed from the chips, it is possible to begin the hydrolysis with an L/S of 1.05 rather than 1.77. The drained liquid may be recycled for reuse (see fig. 27, which gives the water balance).

The decreased L/S results in a significant drop in acid consumption, but the primary process advantage is in the increased concentration of the resulting hydrolysate. If the L/S at initiation of hydrolysis were 1.05, the carbohydrate content of the extract accompanying the residue from the reactor would be around 17% (table 16). It should be possible to recover at least 95% of this material at a carbohydrate concentration exceeding 10%.

The decrease in L/S is dependent on the chip bed depth, which must be shallow enough for the liquid to drain before a significant degree of hydrolysis occurs. In a conventional batch pulping digester, with a bed depth of 20 feet, much of the substrate would be submerged in liquid, most of which could not be removed in the required time interval. There would also be a varying acid concentration vertically throughout the bed, which would have adverse effects on yield and residue quality. It seems the best reactor choice for the first stage is a tubular vessel equipped with a screw auger to move the chips and provided with drainage holes for the escape of liquid during the heatup period. Continuous tubular-type digesters currently used in the pulping industry could perhaps be adapted.

Table 18.—Material balance of first-stage prehydrolysis at reaction conditions of 170° C, L/S = 1.5, pH = 1.45, 11 min

Component	Wood	Solution	Residue
	—% based on OD process wood—		
Total solids	100 (basis)	43.2	62.7
Potential carbohydrates			
Glucose	42.0	3.16	38.6
Mannose	2.37	1.78	.47
Xylose	20.8	16.6	1.38
Uronic	2.84	1.80	.20
Other	2.02	1.57	.28
Total	70.0	24.9	40.9
Acetyl (as HOAc)	6.00	4.50	1.50
Lignin	21.9	3.18	19.2
Extractives	6.67	6.67	—
Ash	.72	.63	.09
Furfural	—	1.51	—
Sulfate (as H_2SO_4)	—	1.73	—
Unaccounted[1]	3.81	—	5.48

[1]Summative analysis can be less or more than 100%.

Processing **Alternatives**

The distribution of the wood components throughout prehydrolysis makes an important point about the process (table 16). The total potential carbohydrates are approximately one-third solubilized and two-thirds in the residue. However, yields in the second stage are known to be very low; not more than 50% of the carbohydrates in the prehydrolysate residue will be obtained from the cellulose hydrolysis. Consequently, the yield of carbohydrates from the second-stage hydrolysis will be only about 20 kg/100 kg OD wood, which is less than that obtained in the prehydrolysate. It is also apparent that the total soluble solids content of the prehydrolysate is considerably greater than that of the second-stage hydrolysate. It is obvious that the manner in which the prehydrolysate is utilized has a major effect on the economic viability of the overall process. Unfortunately, few options exist; only four products seemed worthy of consideration: ethanol, single-cell protein, furfural, and feed molasses.

Ethanol would be preferred since its production is the primary goal. The ethanol production could perhaps be almost doubled by fermentation of the xylose and glucose in the prehydrolysate. Unfortunately, there is no commercial process available to convert xylose to ethanol. Significant progress is being made along this line (Gong et al. 1961; Jeffries 1962; Suihko and Enari 1981), and there seems to be a good possibility of a process being developed; however, the lead time required for commercialization eliminates it from consideration at present.

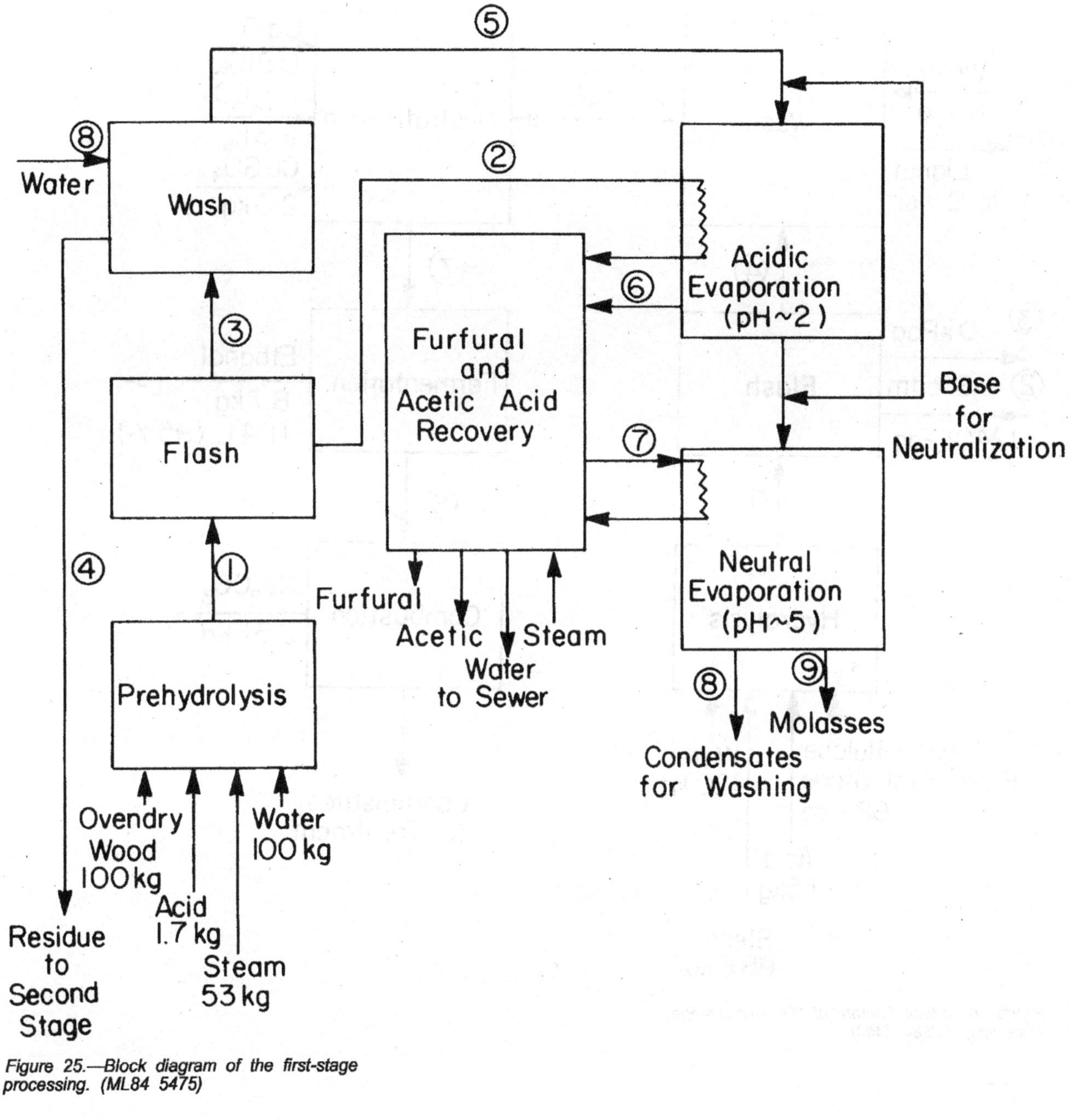

Figure 25.—Block diagram of the first-stage processing. (ML84 5475)

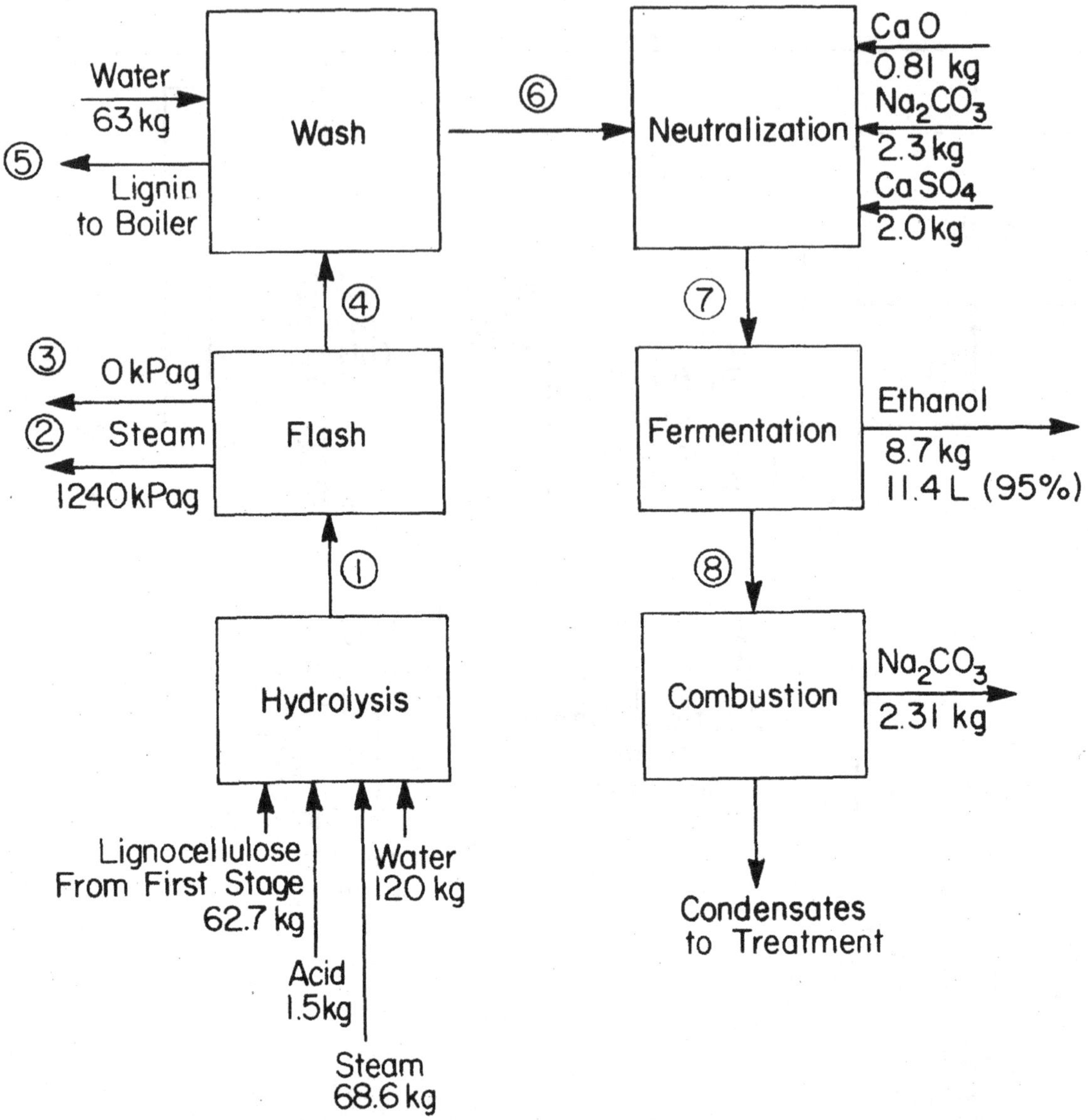

Figure 26.—Block diagram of the second-stage processing. (ML84 5469)

Technology is available for processing the prehydrolysate to yeast and other single-cell protein (Hajny 1981) for animal feedstuffs. In the United States the process would suffer a major disadvantage since it would have to compete with domestic agricultural products generally available in good supply at low cost. The domestic yeast producers have been largely restricted to higher priced, low-volume products for human consumption. Yeast also has a disadvantage, in common with all fermentation products, in leaving an effluent requiring secondary treatment.

The economic picture for the process brightens considerably when furfural is assumed to be the major hemicellulose product, for furfural has a high market value. However, a single hydrolysis plant of the size considered optimal (800 tonne OD wood/day) would result in a 20% increase in the domestic supply of furfural, which is already quite adequate. With several plants, the product might be unmarketable. As described subsequently, the process design includes the recovery of some furfural, but this is less than 20% of the potential that could be produced. Furfural could be considered as a liquid fuel product, but no technology for its use in this manner is available.

At the moment, molasses, marketed as an animal feed, appears to be the preferred way of utilizing the prehydrolysate. This is the basis for subsequent process considerations. Unfortunately the molasses price fluctuates greatly. From mid-1980 to mid-1982, the price of blackstrap molasses fell from $100/tonne to $50/tonne. At the lower price, the production of feed molasses is simply an inexpensive way of disposing of the prehydrolysate. For the process to be viable, it is necessary to find more profitable outlets for the hemicelluloses. The marketing of wood molasses has been difficult and not very successful, but this product should contain less ash and organic impurities than either blackstrap or Masonex (Hajny 1981; Turner 1964).

First-Stage Process Description

Figure 25 presents a flow diagram for the first stage based on molasses as the hemicellulose product. The quantities associated with each numbered stream are listed in table 19.

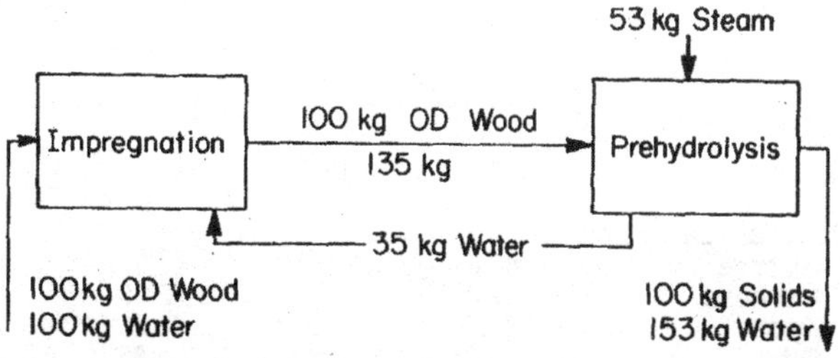

Figure 27.—Impregnation of chips using recycle solution. (ML84 5465)

Stream 1 (fig. 25) is a composite of the solution and residue components listed in table 18 plus 153 kg of water. The water has two sources: 100 kg originate with the incoming wood (50% moisture content) and 53 kg come from the steam required for the hydrolysis. The procedure by which the water-to-wood ratio is kept so low and impregnation accomplished is shown in figure 27. It will be recalled from the previous discussion that the impregnated chips can be easily brought to an L/S of 1.35. This ratio controls the amount of steam that must be added to bring the material to reaction temperature; at this value, 53 kg are necessary. It was also pointed out that on heating in the digester, a large amount of relatively carbohydrate-free liquor drains from the chips. Some of this (35 kg) can be recycled to the impregnator. The water output must equal input, which is 153 kg. This, of course, will vary with the moisture content of the incoming wood within limits imposed by the amount of liquor available for recycling.

Table 19.—Components in the streams of figure 25, first-stage processing

Component	Stream								
	1	2	3	4	5	6	7	8	9
	kg/100 kg OD process wood								
Water	153	18.3	135	120	210	112	40	89	9.2
Potential carbohydrates									
Glucose	41.8	—	41.8	38.8	3.0	—	—	—	3.0
Mannose	2.25	—	2.25	.56	1.69	—	—	—	1.69
Xylose	18.0	—	18.0	2.21	15.8	—	—	—	15.8
Others	1.85	—	1.85	.36	1.49	—	—	—	1.49
Potential acids									
Uronic	2.0	—	2.0	.29	1.71	—	—	—	1.71
Acetic	6.0	.52	5.48	1.70	3.78	2.32	([1])	—	1.46
Lignin	22.4	—	22.4	19.4	3.0	—	—	—	3.0
Extractives	6.67	—	6.67	.33	6.34	—	—	—	6.34
Ash	.72	—	.72	.60[2]	.47	—	—	—	4.20
Furfural	1.51	.55	.96	.06	.90	.85	([1])	.05	
Sulfate (as H_2SO_4)	1.73	—	1.73	.09	1.64	—	—	—	—
Unaccounted	5.48	—	—	5.48	—	—	—	—	—
Total weight	259	19.4	240	184	250	115	42	89	46

Quantities of these components depend on recovery unit design.

[2]Lignocellulose residue has exchange capacity; when washed with hard water, it picks up cations.

The contents of the prehydrolyzer are discharged to the flash unit, releasing vapors (fig. 25, stream 2). This stream contains, in addition to water vapor, considerable amounts of volatile materials formed during digestion-about 40% of the furfural and 12% of the acetic acid as well as some methanol and other minor components such as formic acid introduced with steam recycled from the second stage. The heat in the flashed vapor is utilized in the evaporators and the condensate is sent to the recovery unit. The remaining furfural and acetic acid accompany the wet residue (stream 3) to the washers. Here the solubles are washed. from the residue (95% recovery assumed) with much of the wash water being supplied from evaporator condensate (stream 8). The diluted solution (stream 5) flows to the evaporators, and the residue (stream 4) is passed to the second stage.

The multiple-effect evaporators are operated as two units: the first unit evaporating acidic solution; the second unit, neutral solution. In the first effects, acetic acid as well as furfural is present in the vapors. Condensates of these vapors are the source of stream 6, which is passed to the recovery unit. Between the two units, the solution is neutralized with soluble bases (NaOH or NH_3), which converts the acetic acid to nonvolatile acetate. Thus, the major contaminant in the condensates from the neutral evaporation is the residual furfural that was not released to stream 6. These condensates form stream 8, which is used to wash the residue from the flash unit. The product molasses (stream 9) should be similar to blackstrap molasses in carbohydrate content but contain less ash and nitrogen.

Furfural Recovery

During prehydrolysis, furfural is unavoidably produced. If not recovered, it is a contaminant in the fermentation and a pollutant in the effluent. The yield of furfural during prehydrolysis is a function of time. Under conditions the same as those in table 18, maximum xylose yield occurs at about 6 minutes, at which time the furfural yield is 0.8% based on OD wood, and it is present in the prehydrolysate at a concentration of 0.4% (fig. 28). Furfural can be economically recovered from solutions much less concentrated (Aly and Zacchi 1979; Zacchi and Aly 1979).

Further consideration of furfural production leads to the conclusion that it is economically beneficial to extend the duration of the prehydrolysis and thus produce more furfural. This results from two factors. One is the efficient production of furfural during the prehydrolysis. Prolonging the prehydrolysis time past the maximum xylose yield results in increased furfural but decreased total carbohydrate yield. Both the amount of sugar in solution and the potential sugar in the residue decrease. In a plot of the relationship between furfural gain and carbohydrate loss as the cook is extended past the maximum xylose yield (fig. 29), the early portion of the curve indicates yields greater than stoichiometric yields from xylose. This is because much of the xylose in solution, which reacts to form furfural, is replaced by the solubilization of xylose from the wood. The xylose in the residue is considered to have no value and is not included as part of the carbohydrate loss. The second factor favoring production of more furfural is the small incremental cost incurred to recover the additional amount. The major expense of recovery is the stripping of the dilute solution, and this cost is nearly independent of the furfural concentration.

Optimizing furfural yield requires plant design and other data in more detail than available at this point. It is assumed that it would be near the point of maximum xylan utilization-that is, the point at which the total molar yield of xylose and furfural is a maximum. This is reached at a time of 10.2 minutes when xylan utilization, as described above, is 91.3% and the yield of furfural (on wood) is 1.5%, or about twice that at maximum xylose yield. At this level, furfural production from an 800-tonne/day hydrolysis plant would be 4.2×10^6 kg/year, approximately 4% of current domestic production.

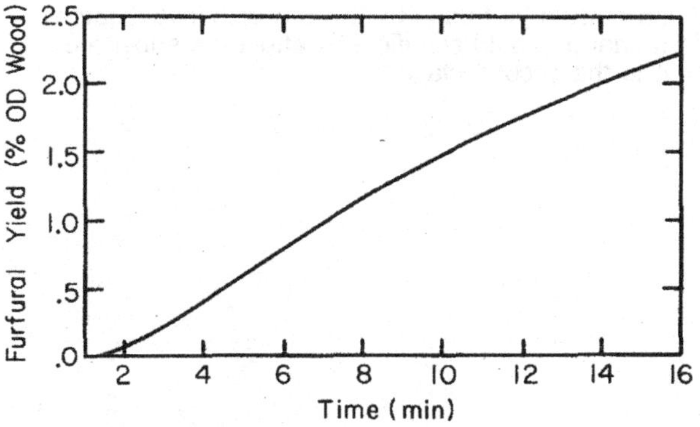

Figure 28.—Calculated furfural yield over time during prehydrolysis at 170° C, pH = 1.42, L/S = 1.5. (ML84 5482)

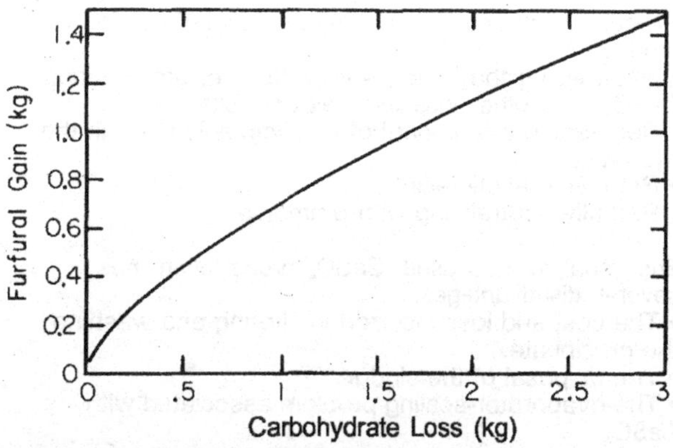

Figure 29.—Calculated furfural gain vs. carbohydrate loss after maximum xylose yield at 170° C, pH = 1.42, L/S = 1.5. (ML84 5477)

Of the furfural from the digester (table 19, fig. 25), approximately 36% (0.55 x 10011.51) is flashed (stream 2), condensed in the evaporators, and passed to recovery as a 2.8% solution. Of the remaining 64%, about 90% is recovered in the evaporator condensates (stream 6) and arrives at the recovery unit as 0.74% solution. Thus, of the furfural produced, more than 90% is transferred to recovery in the two solutions, one 2.8% and the other 0.74%. Both of these solutions also contain methanol and acetic acid and other volatiles, but only volatile components. Almost all of the unrecovered furfural is in stream 8. If stream 8 is used to wash the residue, some additional furfural will be recovered. The key to the economics of recovering furfural from these dilute solutions rests on the double use of steam. Some of the steam shown entering the recovery unit is used to strip the dilute solutions, concentrating the furfural in the vapor (steam) phase. Here the steam is used as a carrier. This furfural-rich steam is then sent to the evaporators (stream 7) where it is condensed, supplying heat to concentrate the carbohydrate solution, and then returned to recovery for further processing. Thus, the steam required for evaporation also serves to supply the largest part of the steam required for furfural recovery.

Acetic Acid Recovery and Molasses Quality
Recovery of the acetic acid becomes economically viable because of its effect on product (molasses) quality and effluent treatment costs.

Wood molasses must compete with blackstrap molasses. Although there are no market standards, it must have approximately the same carbohydrate, nitrogen, and ash content. The ash content of wood molasses has three sources:

- The original ash of the wood.
- The sulfuric acid added for hydrolysis and the base required for its neutralization.
- The base required for neutralization of the acetic and uronic acids liberated on hydrolysis.

Practical methods for reducing the level of ash in the product are:

- Neutralizing the hydrolysate with lime, precipitating $CaSO_4$, and other insoluble calcium salts.
- Decreasing the amount of catalytic acid used in the hydrolysis.
- Removing acetic acid.
- Partially neutralizing with ammonia.

The usual method used, $CaSO_4$ precipitation, has several disadvantages:

- The cost and loss incurred in filtering and washing the precipitate.
- The disposal of the sludge.
- The evaporator-scaling problem associated with $CaSO_4$.

Fortunately, with the process scheme shown in figure 25, the use of lime can be eliminated. The ash content of the molasses can be brought to an acceptably low level using the other procedures-that is, minimizing acid consumption in the prehydrolysis, partially neutralizing with NH,, and removing some of the acetic acid during evaporation. The ash and nitrogen content of the product would be less than that commonly occurring in blackstrap molasses.

It is costly to remove acetic acid from dilute solutions by steam stripping or distillation; solvent extraction methods are much less costly (Hanson 1979; Helsel 1977). However, the better extraction processes involve the use of materials that would be deleterious if present in the molasses. Thus, direct extraction of stream 5 (fig. 25) is not recommended. For this reason, the first evaporators operate under acidic conditions and their condensates (stream 6) containing acetic acid are sent to recovery.

With the process design of figure 25, the only pollutants of the first stage are in the effluent streams emanating from the furfural-acetic acid recovery unit and the neutral evaporators. The latter stream can probably be successfully recycled as wash water, leaving a single stream of environmental concern. Materials in this effluent stream from the recovery unit enter via streams 2 and 6. If carryover is excluded, these streams contain, in addition to water, only volatile compounds, furfural, acetic acid, and minor quantities of methanol, formic acid, acetone, and extractives. The more volatile compounds are isolated during furfural recovery in a combustible stream which may be sent to the boiler. Thus first-stage pollution should be adequately contained by recovering both the furfural and acetic acid.

As indicated in figure 25 and table 19, about 66% (streams 2 and 6 = 2.8 kg) of the acetic acid liberated in the prehydrolysis (streams 2 and 5 = 4.3 kg) goes to the recovery unit, leaving 34% in the product molasses. This acetic acid (2.8 kg) ends up in the still bottoms and condensates from the furfural recovery towers, and could be recovered from these streams. The combined liquid effluents from the recovery unit will contain a total of 170 kg of water and thus, before purification, will have an acetic acid concentration of about 1.5%. At this concentration level, recovery of the acetic acid would be marginally economic at best, but the additional beneficial result on waste treatment makes it desirable. Any formic acid would be removed along with the acetic acid.

The amount of byproduct acetic acid from an 800-tonne/day plant would be 7.5×10^6 kg annually.

Second Stage

Hydrolyzer Type

When this project was initiated, the second-stage hydrolyzer was conceived as a batch digester charged with acid-impregnated chips from the prehydrolysis and heated with direct steam. This method of operation assumes that the original wood chips maintain their integrity through first-stage processing and, when placed in the second-stage digester, have sufficient interstitial space to allow intimate steam contact so that heating is uniform throughout. The presence of fines or compaction of the bed could conceivably result in a nonuniform temperature distribution during the short time interval of the reaction. A batch digester was used successfully in Swedish pilot-plant studies (Cederquist 1954), but this was a small rotating digester. Problems would be anticipated in a stationary vessel of commercial size. Also, the proposed operation of the first stage is considerably different from that used by the Swedish investigators. This change could significantly affect the substrate sent to the second stage.

The decision to produce more furfural by extending prehydrolysis past maximum xylose yield results in approximately doubling the reaction time. The residue produced, although retaining the original chip form, is quite friable and would produce large quantities of fines before reaching the second stage. Since this would make it unsuitable for use in a batch digester, it is recommended that the material leaving the first stage be fiberized before washing; thus the substrate to the second stage (fig. 26) would be a finely divided, fibrous, pulplike material. With such material, the second-stage hydrolyzer would have to be continuous. Commercially available equipment has not been tested at pilot or semicommercial scale, but significant progress is being made in its development (Bender 1979; Church 1981; lotech 1980; McParland et al. 1982). For the subsequent process calculations, it has been assumed that it would be possible to feed a mixture with an L/S of 2. After heating with direct steam, the L/S would rise to 3; and after hydrolysis, at discharge, the L/S would be about 7.

Processing Alternatives
A materials balance for the second-stage hydrolysis was calculated (table 20). The calculation assumes the same reaction conditions as used previously (table 17). It was also assumed that the reaction was stopped just short of the maximum free glucose yield, resulting in a yield of 98% of the maximum predicted. As previously explained, the impurity ratio is significantly decreased by accepting this small yield loss.

About one-half of the charge is solubilized (table 20). The residue is primarily lignin, but, since it contains more than a quarter of the original potential carbohydrates, it has a composition of around 30% carbohydrate. The outstanding property of the soluble solids is the ratio of glucose to other components. Notice that only 61% of the total soluble material is free glucose. The yield of free glucose is only 48.5% of that potentially available. The major nonglucose component is the reversion material, which is also a carbohydrate product and could conceivably be utilized as such. The organic acids are also prominent, and they are of great importance in the economics of processing. The major acidic component is acetic acid, comprising about 60% of the organic acids on a weight basis and about the same on a molar basis. The acetic acid is incompletely removed in the prehydrolysis; 25% is retained by the first-stage solids (table 18). This figure is probably high; some subsequent experiments indicate that it is 16%. Even at the lower value, the acetic acid is a prominent impurity in the second-stage hydrolysate. HMF, which, like formic acid, is toxic in the fermentation, is present in large quantity. Table 20 makes it apparent that the glucose, which is obtained in only moderate yield (48.4%), is accompanied by a large amount of impurities (nonglucose components = 12.0 kg/kg).

Table 20.—Material balance of second-stage hydrolysis at reaction conditions of 230° C, 0.8% (added) H_2SO_4, L/S = 3

Component	Charge[1]	Solution	Residue
	– – – –kg/100 kg OD wood– – – –		
Total solids	62.7	30.7	31.9
Potential carbohydrates			
Glucose	38.6	18.7	10.8
Mannose	.47	.14	
Xylose	1.38	—	—
Reversion material	—	3.1	—
Other	.28	.06	—
Acids			
Acetic	1.50	1.5	—
Levulinic	—	.66	—
Formic	—	.38	—
Sulfuric	—	1.50	—
Klason lignin	19.2	—	20.2
Ash	.47	.45	.09
Furfural	—	.25	—
Hydroxymethylfurfural	—	1.27	—
Unaccounted	5.48	2.50	4.5

[1]Residue from prehydrolysis (see table 16).

In the hydrolysis process we describe here, the unfermentable material has a COD in excess of 100 kg for each tonne of wood charged (table 21). The glucose not utilized in the fermenter must be added to this quantity. Since the efficiency of fermentation could be as low as 90% (Saeman and Andreasen 1954), the total COD load could be near 125 kg/OD tonne of wood. This is 2 to 10 times that encountered In pulping processes. Although there are no standards on allowable discharges for hydrolysis plants, it can be assumed that the standards for pulp production would apply: these are 2-10 kg/OD tonne of wood. Consequently, more than 90% of the pollutants will have to be removed prior to discharge. The cost of doing this by secondary treatment is exorbitantly high, and other means of reducing the COD must be built into the design.

Table 21.—Chemical oxygen demand (COD) for cellulose hydrolysis component

Component	Production on wood basis	COD	
		Component basis	Wood basis
	kg/tonne	*kg/kg*	*kg/tonne*
Monomeric glucose	187.0	1.07	([1])
Reversion material	30.6	1.07	32.7
Acetic acid	15.0	1.07	16.0
Levulinic acid	6.7	1.52	10.1
Formic acid	3.8	.35	1.4
Furfural	2.5	1.67	4.1
Hydroxymethylfurfural	12.7	1.52	19.3
Unaccounted	25.0	1.07	26.7
Total			110.3

[1]For each 1% of the glucose unfermented, add 2.0 kg/tonne.

It would be ideal if the contaminants could be removed as marketable products, as was considered with furfural and acetic acid in the first stage. It might be possible, for example, to recover HMF, levulinic, formic, and acetic acids, and furfural before the fermentation and treat the reversion material enzymatically during the fermentation to convert it to glucose. The removal of the toxic components would probably raise the fermentation efficiency from 90 to 99%, and the reversion material could also be utilized as glucose. The combined effect would increase the alcohol yield by 25%. There would be, in addition, 40 kg/tonne of potential byproducts and a reduction of 78% in the effluent load. Recovering contaminants has a threefold beneficial effect: the recovered material is potentially marketable, its COD contribution is removed, and the efficiency of the fermentation is raised. It is not possible to evaluate processes of this sophistication now because the necessary process information is unavailable. There is also the question of marketing HMF and levulinic acid.

If the impurities cannot be removed, then they should be minimized, but no strategy is available to decrease them to the point at which they could be economically handled by secondary treatment. Two possible alternatives to the more conventional effluent treatment method are producing methane and concentrating & burning. Although methane production would be preferred from an energy viewpoint, combustion is probably superior overall because it has the advantages of complete disposal and the option of future byproduct recovery. We are, thus, led to the simple process shown in figure 26, incorporating combustion disposal into the design.

Two modifications of the design in figure 26 were considered, but neither indicated any improvement. The lignin residue contains approximately 30% of the potential glucose In the charge (table 20). Much of this material could be hydrolyzed by recycling, but the large amount of associated lignin makes this uneconomical. Because the hydrolyzer cannot operate below a minimum US of 3.0, raising the lignin content of the feed results in a higher water-to-cellulose ratio. This increases the heat consumption per unit of sugar production and also decreases the glucose concentration of the hydrolysate.

The possibility of recycling some of the still bottoms was also evaluated. This has the beneficial effect of increasing the yield since the unfermented glucose and reversion material are put back into the system. It also results in a decrease in the effluent load since some of the impurities react and are precipitated as solids. However, the concentration of impurities increases and this has a deleterious effect on the fermentation. The major economic impediment is the large increase in acid consumption. The acidic impurities that must be neutralized for the fermentation must be reacidified for hydrolysis and then neutralized again. Recycling the still bottoms from the fermentation without first removing some of the acid components is probably not advantageous.

Second-Stage Process Description

The flow plan and stream compositions for the second stage of the process are described by figure 26 and table 22. Despite its apparent simplicity, serious questions can be raised about the practicality of almost all the steps. Certain problems require further investigation before plant design can be considered. The design has optimistically been assumed workable so that the overall process can be evaluated. What follows is a brief description of each of the process steps and the possible difficulties associated with each.

As mentioned earlier, it is thought that the hydrolyzer must be a continuous unit. No such equipment has been developed, but prospects seem promising (Bender 1979; Church 1981; Iotech 1980; McParland 1982). Various process designs are possible depending on the outcome of these investigations. The critical economic factor in the development of such equipment is the L/S since it is predominant in determining the energy consumption, not only in the saccharification but also in the subsequent processing. It is almost certain that a 10-15% pulp slurry (L/S = 6-9) could be pumped. Although such dilute slurries have a large heat demand, much of it could be recovered because the slurries are pumpable. The incoming feed to the hydrolyzer could be heated countercurrently with vapors from stepwise flashing of the reactor product. As the L/S is dropped, the feed cannot be handled in this manner and the option of preheating is lost. Heating must be by direct injection of high-pressure steam, and any heat recovered from the product must be utilized elsewhere in the process. It has been assumed here that a feed with L/S ? ?2 can be fed to a reactor and heated with direct steam, which would quickly bring the L/S to 3. The subsequent dissolution of the solid phase would increase the L/S to 7 at discharge.

On discharge from the reactor; the slurry is flashed in two steps. The pressure of the first step is 1,240 kPag, and this steam can be utilized for prehydrolysis. The second step yields steam at atmospheric pressure that can be utilized for evaporation. The vapors from both flash stages contain the contaminants (furfural and acetic and formic acids) shown in table 22 (streams 2 and 3). It should be emphasized that suitable equipment has not been developed to handle lignocellulose in the suggested manner. This is true also in the case of dilute slurries for which the heat-recovery hardware would require development work.

After flashing, the material is passed to washers where the soluble solids are separated from the residue. It is important to operate at high recovery and low dilution. It has been assumed that 95% of the soluble solids could be recovered while limiting the concentration drop from 17.8 to 14.4% total solids. This is very efficient washing and would require sophisticated equipment such as the dewatering presses used in the pulping industry. No data are available on the filtering characteristics of the material. It is very fine but not gelatinous and still retains a large amount of cellulose (table 20). It is probably more amenable to filtration than residues from the percolation process.

For neutralization, it is proposed that two bases be used--lime (CaO) and soda ash (Na_2CO_3). The high concentration of organic acids, all of which have soluble calcium salts, results in a calcium concentration of around 0.5% if the solution is neutralized with lime alone. The high calcium content leads to serious problems in the concentration and combustion of the waste stream. Replacing the

Table 22.—Components in the streams of figure 26, second-stage processing

Component	Stream							
	1	2	3	4	5	6	7	8
	— — — —*kg/100 kg OD process wood*— — — —							
Water	188.6	22.1	27.6	138.9	33	169	169	169
Potential carbo-hydrates								
Glucose	29.5	—	—	29.5	11.7	17.8	17.8	.9
Mannose	.1	—	—	.1	—	.1	.1	—
Reversion	3.1	—	—	3.1	.2	2.9	2.9	2.9
Other	.1	—	—	.1	—	.1	.1	.1
Organic acids[1]								
Acetic	1.5	.23	.27	1.0	.05	.95	.95	.95
Levulinic	.66	—	—	.66	.03	.63	.63	.63
Formic	.38	.03	.05	.30	.02	.3	.3	.3
Lignin	19.9	—	—	19.9	19.9	—	—	—
Ash	.45	—	—	.45	([2])	.45	—	4.0
Furfural	.25	.11	.08	.06	—	.06	.06	.06
Hydroxy-methyl-furfural	1.3	—	—	1.3	.06	1.2	1.2	1.2
Sulfate (H_2SO_4)	1.5	—	—	1.5	.08	1.42	—	—
Unaccounted	7	—	—	7	4.5	2.5	2.5	2.5
Total	253	23	28	208	64	198	198	180

[1]May not be present as free acids.

[2]Unknown-probably less than .005.

calcium ion with sodium ion permits the adoption of technology available from the sulfite pulping Industry. Calcium-free solution can be obtained by the following procedure:

- Neutralize to near pH = 3, precipitating the **sulfate** as $CaSO_4$.
- Filter.
- Ion-exchange to remove residual calcium Ion.
- Raise the pH to 5 with soda ash.

The ion-exchanger can be regenerated with some of the incoming hydrolysate, thus confining the calcium to the first step in the neutralization. A possible difficulty with this procedure is the precipitation of $CaSO_4$ during regeneration. The sulfate content of the final solution is controlled by the pH level following lime neutralization. The higher the pH at this point, the lower the sulfate, but the greater amount of calcium will increase the ion-exchanger load. Notice that removal of calcium is essential if recycling of still bottoms is considered. If the calcium Is not removed, the necessity of reacidification prior to hydrolysis results in a solution saturated with $CaSO_4$, which probably could not be processed.

Several components of the neutralized solution are toxic to yeast, including furfural, hydroxymethylfurfural, and formic acid. Although the untreated material may be fermented using a massive inoculation of adapted yeast, the fermentation efficiency would be much less than the assumed 95%. To attain this high efficiency, pretreatment of some type would be necessary. Detailed discussions of possible methods of pretreatment, nutrient requirements, problems, and performance of the wood-sugar hydrolysate fermentation are available (Hajny 1981; Saeman and Andreasen 1954). A large quantity of the carbohydrate material is not utilized by the yeast; the sum of the monomeric glucose and reversion material leaving the fermentor is equivalent to nearly 25% of the actual alcohol production. Much of this would be available if the toxic materials were removed.

The still bottoms discharged from the fermentation-alcohol recovery unit will contain approximately 5% combustible solids. This is not enough to make the concentration-combustion unit energy self-sufficient, and disposal of the material will increase process heat consumption. There are commercial installations in the pulping industry (Farin 1973) handling solutions of this type that could be readily adapted to this process. Part of the recovered sodium carbonate can be recycled; the amount will depend largely on its sulfate content. The amount and concentration of the condensates resulting from the evaporation will depend on the ultimate design. It is assumed they would present no extraordinary problem.

Material and Energy Balances

A material and energy balance is a method of assessing a process. The input and output of materials for the two-stage hydrolysis process are shown in table 23. Figures 30, 31, and 32 show the distribution of organics, water, and energy. Both the table and figures are based on 100 kg of OD wood entering the first-stage hydrolyzer. Several important assumptions are made in their development.

The wood is assumed to be southern red oak of a composition given in table 1 (sample 3) and a higher heating value of 19.77 MJ/kg (8,500 Btu/lb). The process material delivered from wood preparation is bark free, 50% moisture, reduced to 9.5-mm chips. The bark and some additional fuelwood are sent to the boiler. It is assumed that after impregnation the acid was uniformly distributed throughout the chip so that the yields on prehydrolysis are those shown in table 18. The water content of the prehydrolysate is kept to a minimum by recycling acid solution on the impregnator as shown in figure 27.

Table 23.—Input.output for two-stage hydrolysis

Component	Mass	Value	
	kg/100 kg OD process wood	*$/tonne*	*$*
	INPUT		
Wood (OD)	119[1]	33	3.93
Sulfuric acid	3.3	88	.29
Soda ash	5.3	132	.70
Lime	.8	38.5	.03
Total			4.95
	OUTPUT		
Molasses	47.4	55	2.61
Furfural	1.15	1,450	1.67
Acetic acid	2	580	1.16
Ethanol (100%)	8.7	585	5.09
Subtotal			10.53
Soda ash (byproduct)	2.3	66	.15
Total			10.68

[1]Included 19 kg of wood substance used for boiler fuel.

Conditions of the prehydrolysis are assumed to be the same as those indicated in table 18. The acid required to maintain a pH of 1.45 depends, of course, on the ash content of the wood. This is assumed to be 0.72%, which is the value found for the particular sample used in this study. Residence time in the prehydrolyzer is 12-14 minutes, resulting not in a maximum xylose yield but a maximum summative yield of xylose and furfural. In the second stage, the temperature is 230° C, pH = 1.3, and residence time is 35-40 seconds. As described earlier, to decrease the effluent load, the reaction time is set at less than that required to attain maximum glucose yield. Yields for the various products from the cellulose hydrolyzer are shown in table 20.

It was assumed that 95% of the soluble solids were recovered in each of the washing units. Boiler efficiency was taken as 60%, and the efficiency of the fermentation as 95%.

Figure 30 shows the flow of organics through the process. Notice that 100 kg of OD wood contains 99.3 kg of organic matter since the wood has a 0.7% ash content. Of the total carbon charged (59.1 kg), 37% is in the marketable products, most of which is contained in the molasses; only 7.7% of the total carbon charged is contained in the ethanol. Most of the remainder leaves as carbon dioxide from combustion and fermentation, but 1.3% is present in effluent streams.

The necessity of the ancillary processing steps to reduce the pollution load is clearly evident (fig. 30). if the furfural-acetic acid recovery and evaporation-combustion units were deleted, the waste streams would contain more than 13% of the carbon charged. Furfural and acetic acid removal decreases the effluent load by 90% to about 1.3%. It is probable that the effluent load could be reduced still further because there is a possibility that some of the effluent from the recovery unit could be recycled into the process. The biological oxygen demand (BOD) from an 800-tonne/day plant would be about 16,000 kg/day (from fig. 30).

The water content of the effluent streams is shown (fig. 31) as part of the distribution of process water (cooling water not being considered). The total volume of foul condensate is moderate: for an 800-tonne/day plant, it would be about 2.3×10^3 m^3/day. This results from the low water input to the process, which amounts to only 4.2 m^3/tonne of wood.

The process heat requirements are about 50% greater than the heat available from the lignin residue (fig. 32). The additional energy requirement Is supplied from bark and additional fuelwood. As indicated, about 16% of the material entering wood preparation is sent to the boiler. No electricity needs were considered.

The heat consumed in a particular unit may be supplied either by oxidation of organics or from the enthalpy of entering steam. in some instances-for example, the fermentation-both the potential energy content of organics and the heat content of the steam are consumed. The sum of steam consumed in the furfural-acetic acid recovery and evaporation-combustion units is 279 MJ, which is 40% of the steam generated.

Although more steam enters the second-stage hydrolyzer than any other unit, its energy consumption is the lowest. This is because it is credited with the large amount of low-pressure steam recovered from the second-stage flash unit. The discharge from the reactor is flashed to atmospheric pressure in two steps. The higher pressure steam is used in the first-stage prehydrolysis, and the atmospheric steam is sent to the evaporators. On the basis of available energy, the second-stage hydrolyzer is the major consumer.

The assumptions used in preparing the foregoing information are optimistic. Deductions based on the data in figures 30, 31, and 32 will reflect the best performance that could be expected. Previous sections of this report discussed the technical uncertainties, particularly those of the second stage, and it was apparent that considerable development work remains before the process can be commercialized. Following this development work, it is unlikely that performance as good as that assumed will be obtained.

The suitability of the process as an energy producer can be viewed from several perspectives. Considering it as a source of liquid fuel, the net energy gain as liquid fuel is the energy content of the ethanol product less the energy content of the liquid fuel required to deliver the wood to the plant. Since the latter is 3-5% of the energy content of the delivered material, the maximum energy return as liquid fuel is about 8% of the heating value of the incoming wood. This return would be decreased further by considering the energy required to supply the acid and other process materials, as well as the materials for construction and maintenance of the plant.

Another way to view the energy production is to consider the efficiency of conversion of the energy consumed in the process. The total products-ethanol, furfural, molasses, and acetic acid-have a total heating value of 40.7% of the incoming wood, the ethanol alone 11.1%. Thus, based on the energy consumed (59.3%), the energy yield of the ethanol is 18.7%.

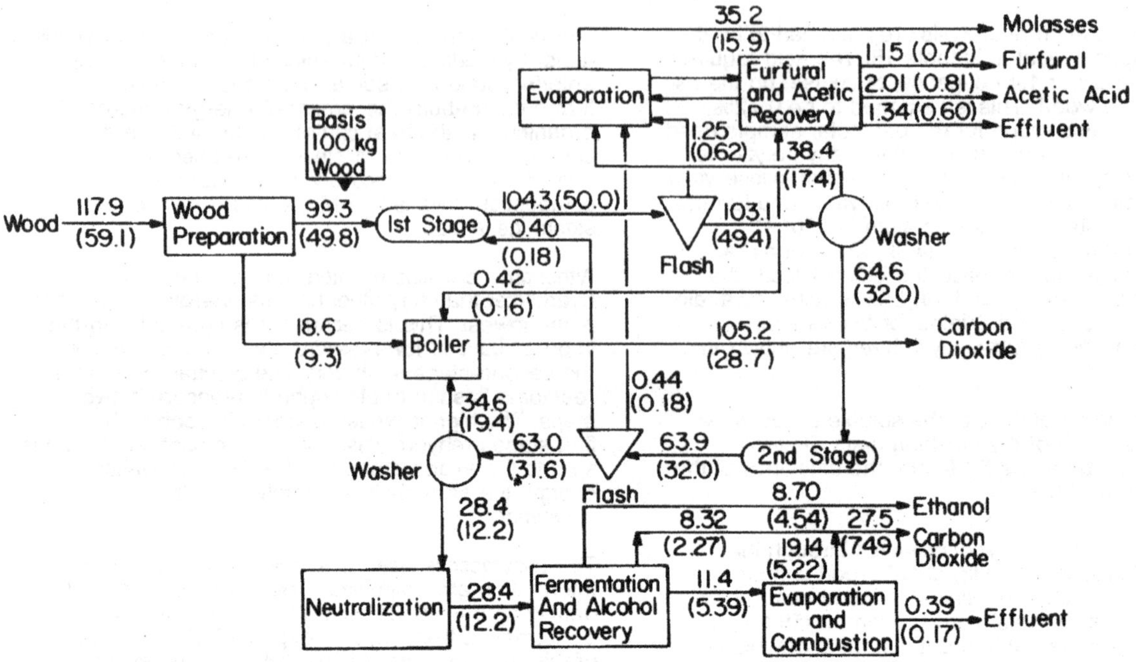

Figure 30.—Distribution of organics through two-stage hydrolysis. Numbers in parentheses are kg of carbon; others are kg of organics. (ML84 5470)

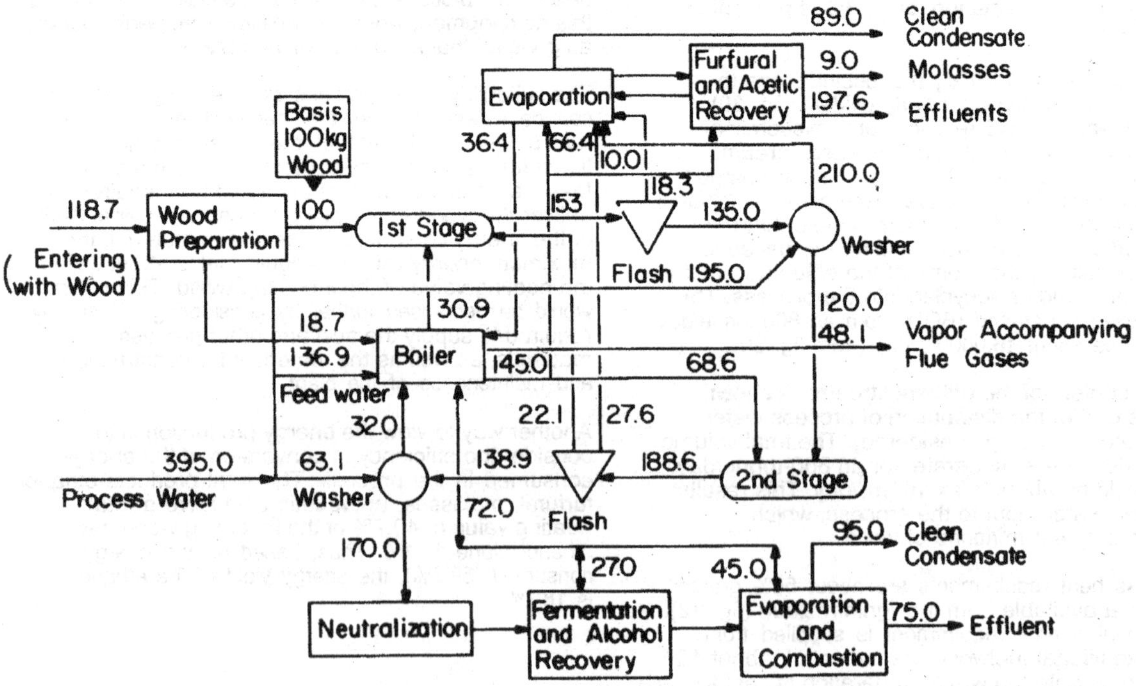

Figure 31.—Distribution of process water through two-stage hydrolysis. All numbers are kg of water. (ML84 5471)

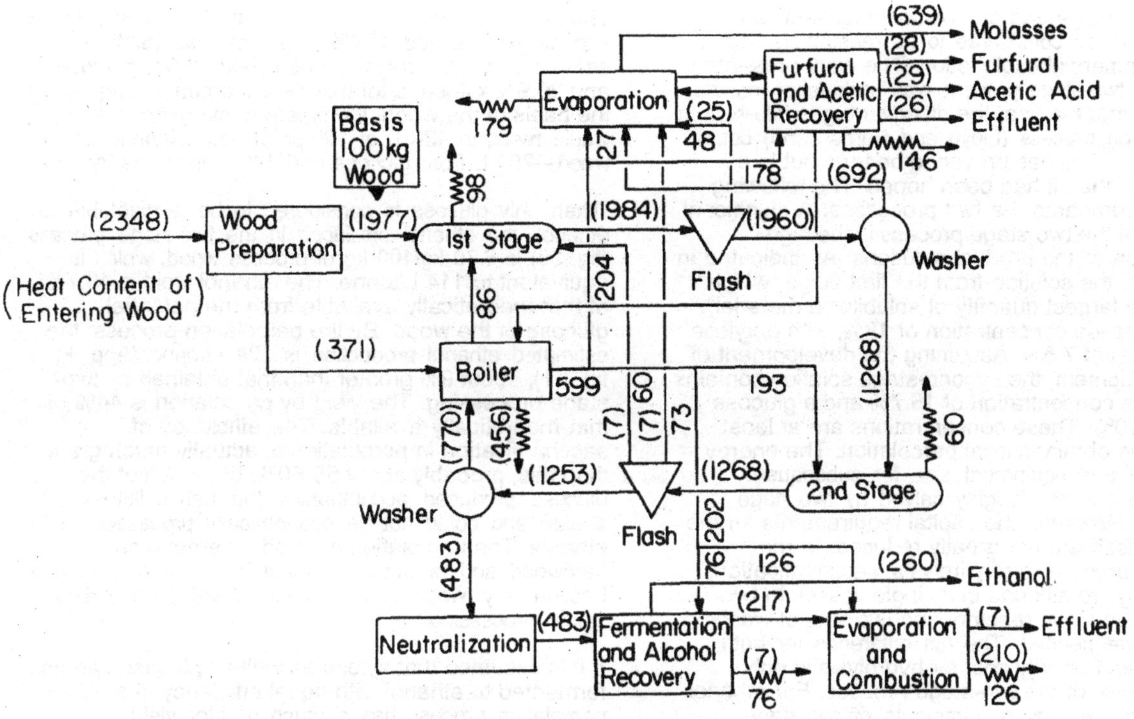

Figure 32.—Distribution of energy through two-stage hydrolysis. Numbers in parentheses are MJ of enthalpy of organics contained in stream; other numbers are MJ of enthalpy of steam. Heat loss from unit is shown as -/\/\/\/-→. *(ML84 5472)*

Comparison with the Percolation Process

Both the two-stage and percolation saccharification processes can be considered to be reasonably near the point of commercial exploitation. The work presented here on the two-stage process was undertaken on the supposition that many of the disadvantages inherent in the percolation process (Lloyd and Harris 1955) could be overcome. This has proved to be true, but to a lesser extent than it had been hoped. The following discussion compares the two processes. A significant advantage of the two-stage process is the high concentration of the product solutions. As indicated in our analysis, the solution from the first stage, which contains the largest quantity of solubilized material, has a total solids concentration of 18%, with a xylose concentration of 7.5%. Assuming the development of suitable equipment, the second-stage solution contains a total solids concentration of 16.7% and a glucose content of 10%. These concentrations are at least double those obtained from percolation. The energy consumption and equipment size for subsequent processing steps are roughly halved by two-stage processing. However, the capital requirements for the hydrolysis itself are not greatly reduced. In the percolation process, prehydrolysis, saccharification, and washing are all done in a single vessel. In two-stage operation, two reactors and two sets of washing equipment are needed. The net difference for both equipment and energy cost for hydrolysis is only slightly in favor of the two-stage process. For ethanol production the energy requirements of two-stage operation should be about 40% less than the percolation process, and equipment cost, about 25% less.

The purity of the solutions generated by the two-stage process is considerably better than that obtained by percolation. The separation between the hemicellulose carbohydrates and the glucose is much better, and the ratio of degradation products to carbohydrates is lower.

The advantages of the two-stage process are offset to some extent by the higher yields of the percolation process. The yield advantage depends both on the type of wood being considered and the subsequent processing of the solution.

Unfortunately, available data on the percolation process relate primarily to softwoods, and yields are given either as total sugars obtained or as recovered ethanol. No data are available on the changing concentration and composition of the product solution. The carbohydrate yields from each of the component fractions of the wood—hemicellulose and cellulose-are unknown. Because of this, the yield comparisons given below are uncertain estimates, but some points for consideration are brought out. Details of the assumptions made in obtaining the yield figures used are given in Appendix F.

The southern red oak used in this study contains 37.8% anhydroglucose and 18.4% anhydroxylose (table 1, sample 3). On hydrolysis, these yield 42.0% glucose and 20.9% xylose, a total of 62.9% potential sugars, on the basis of the wood. Complete conversion to ethanol would result in 420 L of 190-proof ethanol/tonne of wood—280 L from glucose and 140 L from the xylose.

When only glucose is considered to be fermentable to ethanol, the ethanol produced in the two-stage process (fig. 30) is 8.70 kg/100 kg of process wood, which is equivalent to 114 L/tonne. The ethanol yield is 40.7% of that theoretically available from the potential glucose of the wood. By the percolation process, the estimated ethanol production is 124 L/tonne (App. F, fig. F1), about 9% greater than that obtained by two-stage processing. The yield by percolation is 44% of that theoretically available. The efficiency of saccharification in percolation is actually much greater than this, probably about 55-60%, but much of the glucose produced accompanies the hemicellulose stream and could not be economically processed to ethanol. The percolation method, operating on hardwood and producing ethanol from the glucose-rich fraction only, would have a yield advantage over two-stage processing.

If it is assumed that xylose as well as glucose can be fermented to ethanol with equal efficiency, the percolation process has a much greater yield advantage. This is because the xylose-fermenting organism would also utilize glucose, and the poor separation of the sugars is inconsequential. Using this assumption, the percolation-process yield is estimated at 267 L/tonne (App. F, fig. F2) and two-stage operation 234 L/tonne (App. F, fig. F3), a 14% yield advantage for percolation.

Yields for two-stage processing are considerably higher if it is assumed that the reversion material can be converted to ethanol. If this were the case, 133 L/tonne could be obtained solely from the glucose fraction. This increase of 19 L/tonne from the oligomers reverses the ranking and gives two-stage processing a 7.3% advantage under the supposition that xylose is not fermentable to ethanol. If it is assumed that all carbohydrates (xylose, glucose, and reversion products) are fermentable, the percolation process has a 5.5% greater ethanol yield.

Process Research and Development

The economic problems associated with this process are similar to those of any dilute-acid wood conversion process. The general comments given here could equally apply to all dilute-acid saccharification processes. In this section the suggestions are restricted to possibilities for improving the process illustrated in figures 25 and 26. Such improvements could be expected to do no more than raise the process status to where it might be a marginally competitive source of ethanol. It is unlikely that major gains could be made without changing the basic process steps.

If one examines the economic contribution from each of the carbohydrate fractions—hemicellulose and cellulose-the problems associated with each are found to be quite different. The yield from the hemicellulose fraction is high; more than 80% of the potential xylose is recovered. The major shortcoming of hemicellulose utilization is the low value of the product. The molasses contains 70% of the carbon of the marketable products from the process but accounts for only 25% of the total product value. Attaining higher value products from the hemicellulose could significantly improve the economic outlook for the process. The energy and acid consumed in the prehydrolysis, although significant, could probably not be substantially reduced. The situation regarding the utilization of the cellulose fraction is different. Only 44% of the glucose available from the prehydrolysis residue is converted to ethanol. Some loss occurs in washing and fermenting, but the principal problem is the low conversion of cellulose to monomeric glucose in the hydrolysis reaction. High impurity loads, primarily the result of low glucose yields, are also a serious deficiency of the process. The cellulose hydrolysis also consumes large quantities of high-pressure, expensive steam.

The income from the hemicellulose fraction could be increased by producing more furfural. The furfural contributes about 16% to the total product value (table 23). The level of production indicated was set, as previously described, by prolonging digestion past maximum xylose yield to the maximum summative yield of xylose and furfural. Although furfural is generated at decreasing efficiency, probably twice as much could be produced by the same procedure-that is, simply extending the time of prehydrolysis. This would result in some loss in molasses and a small decrease in ethanol production, but the net effect would be favorable, with total product value increasing by about 10%.

Furfural could also be considered as the principal product of the hemicellulose; all of the xylose might be utilized for furfural production. This would require processing of the xylose solution after separating it from the lignocellulose residue. However, serious marketing problems arise, even with the production from a single plant. An 800-tonne/day unit would produce approximately 16 million kg of furfural annually, and this would seriously impact the market. Current U.S. consumption is about 68 million kg/year.

Perhaps the most promising means of improving the utilization of the hemicellulose rests on the success of current research regarding the fermentation of xylose to ethanol (Gong et al. 1981; Jeffries 1982; Suihko and Enari 1981). The ethanol potential of the hemicellulose sugars (120 L/tonne) is slightly more than the estimate of that from the cellulose sugars. Thus, ethanol production could theoretically be doubled. Since this additional ethanol would be produced at the expense of the molasses, the economic outlook would not improve so dramatically. The higher valued ethanol would approximately double the income from the hemicelluloses and increase the gross product value from the process by 25%. Other fermentation products from xylose such as butanediol and itaconic acid (Hajny 1981) could be considered, but these do not seem to have the potential that ethanol has.

In the second-stage hydrolysate, a large quantity of potential glucose is present as reversion material. This is a mixture consisting primarily of [1,6]-linked oligomers of glucose that are not fermented by the yeast. It can be made available for fermentation by hydrolyzing it to the monomer. To do this, two process possibilities exist. The oligomers could be treated enzymatically during the fermentation, or the fermentation bottoms could be recycled to the cellulose hydrolyzer. Unfortunately, both of these options are hampered by the presence of the impurities in solution. Conversion of the reversion material to alcohol, by either method, can be realized only if the impurities are removed.

There are other incentives to remove the degradation products from the second-stage hydrolysate. The major compounds—hydroxymethylfurfural, levulinic acid, and formic acid-are present in substantial quantities even when attempts are made to minimize their production. Their removal and recovery would contribute directly to the economic viabillty of the process, but this would also make indirect contributions. As already pointed out, their removal would open up the possibility of recovering the reversion products. In addition, the efficiency of the alcohol fermentation would be enhanced, and the pollution load from the process would be decreased.

Unlike the prehydrolysis impurities, which can be recovered as marketable products by known processing methods, those of the second stage are not currently marketed, and there is little information regarding means for their recovery. This is an area for research that offers considerable promise for process improvement. Information on the yields of various nonglucose products of acid-catalyzed saccharification and on possible schemes for their recovery is needed.

Much work remains to make the process technically feasible. Suitable equipment must be developed for the reactors of both stages and construction materials selected for these harsh environments. Reliable machinery must also be designed for many of the ancillary process steps, including washing and flashing.

Wood Chemistry Research

Much of the research needed to develop processes for the production of ethanol or other wood-derived chemicals can have broader application. For example, a knowledge of the movement of liquids within impregnated chips during heatup can have important application in some pulping operations, and information on the recovery of furfural and the fermentation of xylose can be used to evaluate byproduct recovery schemes for the sulfite pulping process. In general, we would recommend studies of this nature that have application in the existing industry.

Several studies initiated in this hydrolysis project are incomplete but are being continued:

1. The biochemical conversion of xylose to ethanol and other products. This includes screening of organisms and genetic manipulation of the most promising.
2. The toxicity effects of the contaminants of the second-stage hydrolysate on the *Saccharomyces cerevisiae* Hansen fermentation.
3. The effects of the ash constituents of wood on the first-stage hydrolysis.
4. The kinetics of the release of acetyl and uronic acid groups during hydrolysis with dilute sulfuric acid.
5. The effect of drying on the cellulose portion of the wood.
6. The mechanism of cellulose hydrolysis and studies of possible alternatives to the dilute aqueous sulfuric acid hydrolysis.

Fundamental studies in other areas could be initiated to provide information. These should include work in the general areas of enzymatic hydrolysis of cellulose, solvent delignification, and reactions of carbohydrates in various cellulose solvents.

Summary and Conclusions

This paper presents design information for two-stage, dilute sulfuric acid hydrolysis of southern red oak. Both stages are assumed to be continuous and to be operated with a minimum of liquid. In the first stage, where hemicelluloses are hydrolyzed, acid-impregnated chips, free of all interstitial liquid, are heated with direct steam. A slurry of the washed fibrous material from the first stage is processed in the second stage. Here, under more severe conditions than those in the first-stage reactor, the resistant cellulose is hydrolyzed to glucose.

The report elucidates the physical and chemical aspects of the two reactions and the evolution of a process design. The general strategy was to prepare mathematical models that were then verified or modified by experimental data. Since the motivation for the project was the production of ethanol from hardwoods, the efficiency of the saccharification process was judged with this in mind. Co-products originating from the hemicelluloses were assumed to be animal-feed molasses, furfural, and acetic acid. The experimental work was almost exclusively restricted to investigating the first- and second-stage hydrolysis reactions and did not include studies on the subsequent processing necessary to obtain these products.

For the hemicellulose hydrolysis or prehydrolysis, there were three important findings:

1. When using chips in which the acid solution is uniformly distributed, yields of both xylose and furfural can be satisfactorily predicted from a knowledge of the removal rate of xylose from the wood at the conditions employed. Maximum solubilized xylose yields are increased by increasing either the temperature or acidity of the hydrolysis. However, if the acid solution is poorly distributed throughout the chips, the maximum yield of solubilized xylose is decreased.

2. Of the independent variables-catalyst concentration, liquid-to-solid ratio, and temperature-the catalytic acid concentration is the most difficult to determine and control. Satisfactory prediction of the effective acid concentration was possible if the ash content of the wood and the movement of water and other components during digestion were properly considered.

At low liquid-to-solid ratios, the ash components of the wood significantly reduce the acidity of the system. It was determined that not all of the ash is neutralized during prehydrolysis, but that which is neutralized has a predictable effect on the acidity of the system and consequently also on the xylose and furfural yields.

Heating moist chips with steam results in the movement of much of the solution contained within the chip to the surface. This phenomenon must be considered when calculating acidities.

3. Other interesting particulars of the prehydrolysis were observed:

a. At 170° C, mineral acid is essential to obtaining high yields of solubilized xylose. Autohydrolysis-i.e., hydrolysis with no acid added-results in much lower yields.

b. Dry wood releases more glucose during prehydrolysis than wood that has never been dried.

For the cellulose hydrolysis or saccharification, there were also several findings:

1. It was possible to develop a model to quantitatively predict the products of cellulose hydrolysis. This was done by combining submodels for cellulose hydrolysis, glucose reversion, and glucose decomposition. The required parameters were evaluated from prior data, and additional experiments demonstrated the accuracy of the predictions.

2. The unexpectedly high ash content of the lignocellulose from the first-stage hydrolysis significantly decreased the effective acidity.

3. Reversion materials are principally the (1,6)-linked dimers of glucose and levoglucosan, with the latter becoming predominant at conditions optimum for sugar production. These materials are produced in large quantities at low liquid-to-solid ratios.

4. The products of glucose dehydration—hydroxymethylfurfural, levulinic acid, and formic acid-are prominent in the hydrolysate. The amount and relative proportions of each are highly dependent on the reaction time.

Process considerations have led to several conclusions:

1. Furfural and a portion of the acetic acid should be recovered as by-products, because of both economics and efficiency of processing. Income from the furfural could contribute significantly to the economic viability of the process.

2. The best existing design for utilizing the hydrolysate of the second stage is one in which all products other than glucose are concentrated and burned. There is not sufficient information regarding the recovery and utilization of the nonglucose products to develop alternate designs.

3. The two-stage saccharification process for ethanol production is technically ready for commercial exploitation, but some equipment testing and evaluation are necessary first. However, our evaluation of the process at its present stage of development shows that it utilizes the potential energy of the wood inefficiently and is economically unattractive. Future increases in the costs of fossil fuel and agricultural raw materials, such as corn, could result in the process becoming economical.

4. Future research could improve the potential of the process. The major problem areas are:

a. Improved utilization of the pentose sugars, which might be realized by fermentation to ethanol.

b. Increasing yields of marketable products from the cellulose fraction. This could be accomplished by developing saccharification procedures giving higher yields of glucose or by utilizing the large quantities of products other than glucose.

c. Development and evaluation of processing equipment.

Literature Cited

Aly, G.; Zacchi, G. Simulation and design of distillation units for treatment of sulfite pulping condensates to recover methanol and furfural. Part II: Applicability of multiple-effect distillation using live steam. Canadian Journal of Chemical Engineering. 57(6): 316-320; 1979.

American Society for Testing and Materials. Standard test method for ash in wood. ASTM D 1102-56. Philadelphia, PA: ASTM; 1981.

American Society for Testing and Materials. Standard method for chromatographic analysis of chemically refined cellulose. ASTM D 1915-63. Philadelphia, PA: ASTM; 1981.

BeMiller, J. N. Acid hydrolysis and other lytic reactions of starch. Chapter XX. In: Whistler, R. L.; Paschall, E. F., eds. Starch: Chemistry and technology, Vol. 1. New York: Academic Press; 1965.

Bender, R. Method of treating lignocellulose materials to produce ruminant feed. U.S. Patent 4,136,207; January 23, 1979.

Browning, B. L. Apparatus for determination of polyuronide carboxyl. Tappi. 32(3): 119; 1949.

Cederquist, K. N. Some remarks on wood hydrolyzation. In: The production and use of power alcohol in Asia and the Far East: Report of seminar [1952 October 23-November 6; Lucknow, India]. New York: United Nations; 1954: 193-197.

Dunaway, J. W. Levulinic acid in the acid-decomposition products of glucose. M.S. thesis. Madison, WI: University of Wisconsin, Department of Chemistry; 1950. 43 p.

Church, J. A. Continuous high-solids acid hydrolysis of biomass in a 1-1/2-in. plug flow reactor. Industrial and Engineering Chemistry Product Research and Development. 20(2): 371-378; 1981.

Farin, W. G. Economical approaches to sulfite recovery. Tappi. 56(9): 69-72; 1973.

Feather, M. S.; Harris, J. F. Dehydration reactions of carbohydrates. Advances in Carbohydrate Chemistry and Biochemistry. 28: 161-224; 1973.

Gong, C.; Ladisch, M. R.; Tsao, G. T. Production of ethanol from wood hemicellulose hydrolyzates by a xylose-fermenting yeast mutant, *Candida* sp. XF 217. Biotechnology Letters. 3(11): 657-662; 1981.

Hajny, G. J. Biological utilization of wood for production of chemicals and foodstuffs. Res. Pap. FPL 385. Madison, WI: U.S. Department of Agriculture, Forest Service, Forest Products Laboratory; 1981. 64 p.

Hanson, C. Solvent extraction-An economically competitive process. Chemical Engineering Journal. 86(10): 83-87; 1979.

Harris, J. F. Acid hydrolysis and dehydration reactions for utilizing plant carbohydrates. In: Timell, T. E., ed. Proceedings of 8th Cellulose Conference. I. Wood Chemicals-A Future Challenge [1975 May 19-23; Syracuse, NY]. Applied Polymer Symposium No. 28. New York: John Wiley and Sons; 1975: 131-144.

Harris, J. F.; Saeman, J. F.; Locke, E. G. Wood as a chemical raw material. Chapter 11. In: Browning, B. L., ed. The chemistry of wood. New York: Interscience Publishers; 1963: 535-585.

Helsel, R. W. Removing carboxylic acids from aqueous wastes. Chemical Engineering Progress. 73(5): 55-59; 1977.

Iotech Corporation. Optimization of steam explosion pretreatment. Final Rep., DOE Contract No. DE-AC02-79ETZ3050. Ottawa, Ont., Canada: Iotech Corp.; 1980. 360 p.

Jeffries, T. W. A comparison of *Candida tropicalis* and *Pachysolen tannophilus* for conversion of xylose to ethanol. In: Scott, C. D., ed. Proceedings of 4th Symposium on Biotechnology in Energy Production and Conservation [1982 May 11-14; Gatlinburg, TN]. Biotechnology and Bioengineering Symposium No. 12. New York: John Wiley and Sons; 1982: 103-110.

Kerr, R. W. Chemistry and industry of starch; starch sugars and related compounds. New York: Academic Press; 1944. 472 p.

Kirby, A. M., Jr. Kinetics of consecutive reactions involved in wood saccharification. M.S. thesis. Madison, WI: University of Wisconsin, Department of Chemical Engineering; 1948.

Lloyd, R. A.; Harris, J. F. Wood hydrolysis for sugar production. Rep. 2029. Madison, WI: U.S. Department of Agriculture, Forest Service, Forest Products Laboratory; 1955.

Lundquist, K.; Simonson, R.; Tingsvik, K. Studies on lignin carbohydrate linkages in milled wood lignin preparations. Svensk Papperstidning. 83: 452-454; 1980.

McKibbins, S. W. Kinetics of the acid catalyzed conversion of glucose to 5-hydroxymethyl-2-furaldehyde and levulinic acid. Ph. D. thesis. Madison, WI: University of Wisconsin, Department of Chemical Engineering; 1958. 252 p.

McKibbins, S. W.; Harris, J. F.; Saeman, J. F.; Neill, W. K. Kinetics of the acid-catalyzed conversion of glucose to 5-hydroxymethyl-2-furaldehyde and levulinic acid. Forest Products Journal. 12(1): 17-23; 1962.

McParland, J. J.; Grethlein, H. E.; Converse, A. O. Kinetics of acid hydrolysis of corn stover. Solar Energy. 28(1): 55-63; 1982.

Millett, M. S.; Moore, W. E.; Saeman, J. F. Preparation and properties of some hydrocelluloses. Industrial and Engineering Chemistry. 46(7):1493-1509; 1954.

Minor, J. L. Nonfermentable glucose-containing products formed from glucose under cellulose hydrolysis conditions. In: Sarko, A., ed. Proceedings of 9th Cellulose Conference. II. Symposium on Cellulose and Wood as Future Chemical Feedstocks and Sources of Energy, and General Papers [1982 May 24-27; Syracuse NY]. New York: John Wiley and Sons; 1983: 617-629.

Nelson, N. A photometric adaptation of the Somogyi method for the determination of glucose. Journal of Biological Chemistry. 153: 375; 1944.

Palce, M. G.; Jurasek, L.; Desrochers, M. Simplified analysis of wood sugars. Tappi. 65(7): 103-106; 1982.

Pettersen, R. C.; Schwandt, V. H.; Effland, M. J. An analysis of the wood sugar assay using high performance liquid chromatography (HPLC). Journal of Chromatographic Science. 22(11): 478-484; 1984.

Root, D. F. Kinetics of the acid catalyzed conversion of xylose to furfural. Ph. D. thesis, Madison, WI: University of Wisconsin, Department of Chemical Engineering; 1956. 248 p.

Root, D. F.; Saeman, J. F.; Harris, J. F.; Neill, W. K. Kinetics of the acid-catalyzed conversion of xylose to furfural. Forest Products Journal. 9(5): 158-164; 1959.

Rydholm, S. A. Pulping processes. New York: Interscience Publishers; 1965. 1269 p.

Saeman, J. F. Kinetics of wood saccharification. Industrial and Engineering Chemistry. 37(1): 43-52; 1945.

Saeman, J. F.; Andreasen, A. A. Production of alcohol from wood waste. In: Underkofler, L. A.; Hickey, R. J., eds. Industrial fermentations, Vol. I. New York: Chemical Publishing Co., Inc.; 1954: 136-171.

Saeman, J. F.; Kirby, A. M., Jr.; Young, E. P.; Millett, M. A. Rapid high temperature hydrolysis of cellulose. Res. and Mark. Act Stud. Rep. 30. Madison, WI: U.S. Department of Agriculture, Forest Service, Forest Products Laboratory; 1950.

Saeman, J. F.; Moore, W. E.; Mitchell, R. L.; Millett, M. A. Techniques for the determination of pulp constituents by quantitative paper chromatography. Tappi. 37(8): 336; 1954.

Schaffer, P. A.; Somogyi, M. J. Copper-iodometric reagents for sugar determination. Journal. of Biological Chemistry. 100: 695; 1933.

Scott, R. W. Calorimetric determination of hexuronic acids in plant materials. Analytical Chemistry. 51(7): 936-941; 1979.

Scott, R. W. Combined determinations of glucose, mannose, and xylose by spectrophotometry. Analytical Chemistry. 48(13): 1919-1922; 1976.

Scott, R. W.; Green, J. Quantitative spectrometric determination specific for mannose. Analytical Chemistry. 46(4): 594-596; 1974.

Smith, P. C.; Grethlein, H. E.; Converse, A. O. Glucose decomposition at high temperature, mild acid and short residence times. Solar Energy. 28(1): 41-48; 1982.

Sookne, A. M.; Harris, M. The relation of cation exchange to the acidic properties of cellulose. Textile Research Journal. 10: 405-419; 1940.

Spriggs, A. S. Studies on the composition of hydrolysates and glucose reversion mixtures. Ph. D. thesis. St. Louis, MO: Washington University, Department of Chemistry; 1954. 96 p.

Springer, E. L. Hydrolysis of aspenwood with aqueous Solutions of hydrochloric acid. Tappi. 49(3): 102-106; 1986.

Springer, E. L. Rate studies of the hydrotropic delignification of aspenwood. Ph. D. thesis. Madison, WI: University of Wisconsin, Department of Chemical Engineering; 1961. 217 p.

Springer, E. L.; Harris, J. F. Prehydrolysis of aspenwood with water and with aqueous sulfuric acid. Svensk Papperstidning. 85(15): R152-154; 1982.

Springer, E. L.; Harris, J. F.; Neill, W. K. Rate studies of the hydrotropic delignification of aspenwood. Tappi. 46(9): 551-555; 1963.

Springer, E. L.; Zoch, L. L. Hydrolysis of xylan In different species of hardwoods. Tappi. 51(5): 214-218; 1968.

Sroczyński, A.: Boruch, M. Condensation of D-glucose in acid solutions. Staerke. 16(7): 215; 1964.

Suihko, M. L.; Enari, T. M. The production of ethanol from D-glucose and o-xylose by different ***Fusarium*** strains. Biotechnology Letters. 3(12): 723-728; 1981.

Technical Association of the Pulp and Paper Industry. Preparation of wood for chemical analysis. T264 OM-82. Atlanta, GA: TAPPI. 1982.

Technical Association of the Pulp and Paper Industry. Acid insoluble lignin in wood and pulp. T222 OS-74. Atlanta, GA: TAPPI; 1982.

Technical Association of the Pulp and Paper Industry. Ash in wood and pulp. T211 OM-80. Atlanta, GA: TAPPI; 1982.

Turner, H. D. Feed molasses from the Masonite process. Forest Products Journal. 14(7): 282-284; 1964.

Wentz, F. E.; Marcy, A. D.; Gray, M. J. Analysis of wood sugars in pulp and paper industry samples by HPLC. Journal of Chromatographic Science. 20: 349-352; 1982.

Wiesenberger, E. The microanalytical determination of C-methyl and acetyl groups. Macromolecular Chemistry. 33: 51-69; 1947.

Young, E. P. The effect of temperature on the yield of sugar from cellulose. Res. and Mark. Act Stud. Rep. 15. Madison, WI: US. Department of Agriculture, Forest Service, Forest Products Laboratory; 1949.

Zacchi, G.; Aly, G. Simulation and design of distillation units for treatment of sulfite pulping condensates to recover methanol and furfural. Part I: Incorporation with an evaporation unit and use of secondary steam. Canadian Journal of Chemical Engineering. 57(6):311-315; 1979.

Ziminski, R. D. Conversion of cellulose waste to low-cost ethanol. AIChE Symposium Series. 79(223): 8-14; 1983.

Appendix A
Acidity Calculations

I. Calculation of Hydrogen Ion Concentration $[H^+]$

Basis:

100 kg OD wood

Given:

Concentration of added acid = CA (% H_2SO_4)

Liquid-to-solid ratio (L/S) = LS (kg/kg)

Neutralizing capacity of wood = EQW (eq/kg OD wood)

Neutralizing capacity of residue = EQR (eq/kg OD wood)

Then:

Cations neutralized = EQ = EQW - EQR (eq/kg OD wood)

Volume of solution = 100 x LS (liters, assuming a density of 1 g/ml)

Sum of sulfate and bisulfate in solution = SSS[1] = CA/9.8 (mols/L)

Cations in solution = CS[1] = EQ/LS (eq/L)

Sulfur Balance:*

$SSS = [HSO_4^-] + [SO_4^=]$

Charge Balance:

$CS = [HSO_4^-] + 2[SO_4^=] + [OH-] - [H^+]$

Equilibrium Constants:

$FKW = [H^+][OH^-]$

$FKA = [H^+][SO_4^=]/[HSO_4^-]$

The last four equations reduce to the cubic equation:

$A(1)[H^+]^3 + A(2)[H^+]^2 + A(3)[H^+] + A(4) = 0$

where

A(1) = 1

A(2) = CS + FKA – SSS

A(3) = CS x FKA – FKW – 2 x SSS x FKA

A(4) = – FKW x FKA

With numerical values for CS, SSS, FKW, and FKA, the coefficients may be evaluated and the cubic equation solved for its one real positive root which is the value of $[H^+]$. The values of FKA and FKW depend on temperature. At 25° C, they have the values 1.2×10^{-2} and 1.0×10^{-14}, respectively; at 170° C, 1.6×10^{-4} and 3.1×10^{-12}.

II. Calculation of Necessary Concentration of Added Acid to Obtain a Particular $[H^+]$

Basis:

100 kg OD wood

Given:

Liquid-to-solid ratio (L/S) = LS (kg/kg)

Hydrogen ion concentration = $[H^+]$ (mols/L)

Neutralizing capacity of wood = EQW (eq/kg OD wood)

Neutralizing capacity of residue = EQR (eq/kg OD wood)

Then:

Cations neutralized = EQ = EQW – EQR (eq/kg OD wood)

Volume of solution = 100 x LS (liters, assuming a density of 1 g/ml)

Cations in solution = CS = EQ/LS (eq/L) assuming no $CaSO_4$ precipitation

Charge Balance:

$CS - [OH^-] + [H^+] = [HSO_4^-] + 2[SO_4^=]$

Equilibrium Constants:

$FKW = [OH^-][H^+]$

$FKA = [H^+][SO_4^=]/[HSO_4^-]$

From which,

$[SO_4^=] = (CS - FKW/[H^+] + [H^+])/([H+]/FKA + 2)$

$[HSO_4^-] = [SO_4^=][H^+]/FKA$

Thus,

$SSS = [SO_4^=] + [HSO_4^-]$, and CA = 9.8 x SSS

---

[1]It is possible that calcium ions, extracted from the wood, might precipitate as calcium sulfate. This would be a factor in calculating [H+], but it is difficult to bring in quantitatively. However, it was concluded that, at the conditions of prehydrolysis, precipitation would not occur because of the low concentration of sulfate ion. The secondary ionization constant for sulfuric acid decreases rapidly with temperature, and at 170° C and pH = 1.5, only about 0.5% of the acid is present as sulfate ion.

[2]Conventional concentration units of mols/L are used–i.e., [x] = concentration of x in mols/L.

Appendix B
Calculation of Furfural Yields During Xylan Hydrolysis

The differential production of furfural from xylose via the mechanism

$$\text{Xylose} \xrightarrow{k} \text{Intermediates} \xrightarrow{k_1} \text{Furfural} \xrightarrow{k_2} \text{Decomposition Products}$$

Intermediates + Furfural $\xrightarrow{k_3}$ Decomposition Products

was approximated by Root (1956) using the equation:

$$\frac{dF}{dt} = k \times (X - a \times F - b \times F \times X/XO)$$

where

t = time

k = reaction rate constant for xylose disappearance

X and F = molar concentrations of xylose and furfural

a = a(T,XO) (see Root 1956), where T = temperature

b = b(T,XO) (see Root 1956)

X0 = initial molar concentration of the xylose solution

The functions a and b were evaluated from -experimental data, resulting in an excellent correlation. An integrated solution of the above equation was also presented, but only the differential form was used here.

Root's experimental procedure was to load small glass ampoules with solutions of various acid and xylose concentrations, react them at various temperatures for timed intervals, and measure the furfural yield. The quantity XO was a clearly defined, experimental variable included in the empirical correlation. In applying the correlation to the prehydrolysis, a problem arises as to the value to use for XO. The correlation is based on experiments in which the xylose concentration, initially XO, decreases exponentially with time. In the prehydrolysis, the xylose concentration is initially zero, rises to a maximum, and then decreases. Obviously, the relationship between the intermediate concentration and xylose concentration is very different from that present in Root's (1958) study.

Furfural production in the prehydrolysis was obtained by simultaneous numerical integration of the xylan removal rate curve, the xylose degradation rate equation, and the furfural rate equation above. The, furfural production rate, dF/dt, was evaluated at each instant in time by assuming two values for XO. One value used for XO was the actual xylose concentration at the moment; the second was the hypothetical xylose concentration assuming the total of all the xylose released up to that time to be present. Using these two values of XO, two values for each of a and b can be obtained. Thus, using all combinations of a and b, four values of dF/dt can be calculated. Of these four the minimum value was chosen.

Appendix C
Analysis of Glucose Reversion Data

The following procedure was used to extract values of the equilibrium constants, CD and CL, from the experimental data (table 10) obtained by the procedure described in the text. It was assumed that the system was at equilibrium (fig. 11) and thus:

D $= k_2G^2/k_3$

L $= k_4G/k_5$

where

D = disaccharide (mols/L)

L = levoglucosan (mols/L)

G = free glucose (mols/L)

Two values for the combined glucose can be obtained from the data (table 10):
1. Glucose after hydrolysis by the reducing method (col. 7) less the free glucose accompanying the reversion material (col. 4).
2. Glucose after hydrolysis by the oxidase method (col. 5) less (col. 4).
The average of these two values is the combined glucose or reversion material, CG:
CG = ((col. 7) + (col. 5) – 2 x (col. 4))/2 (mg/mL)
The total glucose, TG, is the sum of the free, G, and combined, CG, glucose:

TG = (col. 3) + CG (mg/ml)

The specific reducing power, RP, of the reversion material:

RP = ((col. 6) – (col. 4))/CG (mg glucose/mg combined glucose).

Assume that the reducing power of a mol of disaccharide is the same as that of a mol of glucose and that levoglucosan has no reducing power; then:

D = RP x CG/180 (mols/L)

and, since 2D + L = CG/180, then:

L = (1 – 2 x RP) x CG/180 (mols/L)

Thus the equilibrium constants, CD and CL, can be evaluated,

$CD = k_2/k_3 = D/G^2$ (L/mol)

$CL = k_4/k_5 = L/G$

where

G = free glucose

= (col. 3)/180 (mol/L)

The fraction of combined glucose present as levoglucosan is:

L/(2D + L) = (1 – 2 x RP)

Appendix D
Relationships Between Disappearance Rates of Reducing Power, Free Glucose, and Total Glucose

Define:

TG = total glucose (mol/L)

G = free glucose (mol/L)

D = disaccharide (mol/L)

L = levoglucosan (mol/L)

R = solution reducing power (mol glucose/L)

RPD = specific reducing power of disaccharide (mol glucose/mol disaccharide)

CD = disaccharide equilibrium constant

CL = levoglucosan equilibrium constant

Assume reversion equilibrium established, then:

D = CD x G^2

L = CL x G, and

R = G + RPD x D

If RPD = 1.0, R = G + D

(dR/dt) = (dG/dt) + (dD/dt)

= (1 + 2 x CD x G) x (dG/dt)

TG = G + 2 x D + L
= G + 2 x CD x G^2 + CL x G

Thus, G = (-B + SQ)/2

where

B = (1 + CL)/(2 x CD), and

SQ = $\sqrt{B \times B + 2 \times TG/CD}$

(dG/dt) = (d(SQ)/dt)/2

= (d(TG)/dt)/(2 x CD x SQ)

(dR/dt) = (1 + CD x (-B + SQ)) x (dG/dt)
= (1 + CD x (-B x SQ)) x (d(TG)/dt)/(2 x CD x SQ)

Define:

F1 = 1 + CD x(-B + SQ)

F2 = F1/(2 x CD x SQ)

Then:

(dR/dt) = F1 x (dG/dt)
= F2 x (d(TG)/dt)

The function F1 increases from 1.0 to 1.10, and F2 decreases from 0.90 to 0.84 as the total concentration changes from 0 to 20% glucose.

Appendix E
Calculation of Glucose Yields
(From Fig. 16)

Basis:
100 g lignocellulose (LC)

Define:

G	= free glucose (mol/L)
L	= levoglucosan (mol/L)
D	= disaccharides (mol/L)
TG	= total glucose, free + combined
	= G + 2 x D + L (mol/L)
R	= reducing power of solution
	= G + D (mol/L)
v	= volume of hydrolyzing solution at 25° C (L)
t	= time (min)
FKC	= cellulose weight loss rate constant (min^{-1})
FKR	= reducing power rate constant (min^{-1})

Given:
Total cellulose in lignocellulose = CL (%)

Cellulose resistant to hydrolysis = CR (% of total cellulose)

Liquid-solid ratio = LS (kg/kg OD lignocellulose)

Concentration of added acid = CA (%)

Temperature = T (°C)

Neutralizing capacity of lignocellulose = EQW (eq/kg OD lignocellulose)

Neutralizing capacity of hydrolyzed residue = EQR (eq/kg OD lignocellulose)

Equilibrium constant for disaccharides;
$D = CDxG^2 = CD$

Equilibrium constant for levoglucosan; L = CLxG = CL

Numerically integrate:

$$d(TG)/dt = FKC \times CL \times CR \times \exp(-FKC \times t)/(16200 \times V) - FKR \times R/F2 \quad (E1)$$

$$(TG)_{t=0} = CLx(100 - CR)/(16200 \times VO) \quad (E2)$$

Equation (EI) assumes that k_1 and k_2>>kg (fig. 16) and that L and D are in complete equilibrium with G. It is also assumed that the easily hydrolyzed cellulose is instantaneously converted to glucose giving the boundary condition of eq. (E2). VO is the volume of the solution (associated with 100 g of lignocellulose) containing this amount of glucose-that is, the glucose liberated from the easily hydrolyzed cellulose.

The first term on the right of eq. (E1) is the production of total glucose from the resistant cellulose; the second term is the destruction of total glucose. The rate constant for cellulose hydrolysis is calculated from the equation:

$$FKC = 2.80 \times 10^{20} \times (CH)^{1.218} \times \exp(-21600/(T + 273.1)) \quad (E3)$$

Here CH is the H-ion molarity of the solution after neutralization of the ash constituents but before release of the easily hydrolyzable cellulose. It is calculated by the procedure in Appendix A using the given values of CA, LS, EQW, and EQR. Consistent with the handling of the experimental data on which eq. (E3) is based, CH does not vary as the reaction proceeds and thus, at isothermal conditions, FKC has a constant value.

The volumetric factor, V, is a function only of glucose concentration and does not vary with temperature. The volume of the reacting solution does vary with temperature, but this effect is implicitly included in the correlation of the rate constant, FKC, and the equilibrium constants, CD and CL. The factor V varies throughout the reaction period and is related to TG through the density (DEN, g/L) by:

$$V = LSx\ 100/(DEN \times 1000 - GT \times 180) \quad (E4)$$

The solution densities are assumed to be those of glucose solutions of the same concentration at an ambient temperature of 20° C. Thus eq. (E4) expresses V as a function of TG.

The second term on the left of eq. (E1) accounts for the disappearance of total glucose, TG, from the solution. The relationship between the differentials for R and TG has been shown to be:

$$d(TG)/dt = (dR/dt)/F2 \quad (E5)$$

and since dR/dt = FKR x R, the rate of total glucose disappearance, the second term, on the left of eq. (E1), is FKR x R/F2. Each of the factors FKR, R, and F2 is related to TG:

R is the reducing power of the solution-that is, R = G + D. It is converted to a function of TG as shown in Appendix D. Using that nomenclature:

$$R = G + CD \times G^2 \quad (E6)$$

and

$$G = (-B + SQ)/2 \quad \text{(E7)}$$

F2 is related to TG by the expression derived from Appendix D:

$$F2 = (1 + CD \times (-B + SQ))/(2 \times CD \times SQ)$$

The parameters CD and CL are assumed to be exponential functions of temperature, the expressions having been derived from reversion data taken at 180° and 230° C:

$$CD = 1.127 \times 10^{-5} \times \exp(4272/(T + 273.1)) \quad \text{(E8)}$$

$$CL = 4.758 \times 10^{4} \times \exp(-6471/(T + 273.1)) \quad \text{(E9)}$$

FKR, the rate constant for the disappearance of reducing power, is obtained from McKibbins' correlation (1958). The inputs required are T, acid normality (N), and glucose concentration. The glucose concentration is assumed to be GT. The acid normality varies with the solution volume which, as described above, is dependent on GT. It is determined as follows.

The initial acid normality is established from CH, the H-ion molarity used above to calculate FKC. This molarity corresponds to a particular sulfuric acid normality, which may be determined from CH by the procedure given in Appendix A, assuming EQ = O. However, at the pH level considered here, it can be assumed that only the first hydrogen on sulfuric acid is effective and the initial normality, prior to any hydrolysis, is simply twice the H-ion molarity. The normality throughout the reaction is inversely related to the volume change, and thus to TG by the equation:

$$N = 2 \times CH \times (DEN \times 1000 - GT \times 180)/1000 \quad \text{(E10)}$$

It is assumed in eq. (E10) that the density of water at ambient temperature is 1 g/cm^3.

With all its parameters related to TG, the differential eq. (E1) with its boundary condition can be numerically integrated to obtain TG as a function of time. This is most readily done using concentration units of mol/L and subsequently using the volumetric relationships, converting to percentage yields.

Appendix F
Comparison of Ethanol Yields from Percolation and Two-Stage Processes

Estimation of Saccharification Efficiency by Percolation

We can suppose that Douglas-fir, with a composition of 45% glucan and 10.8% mannan, gives an ethanol yield of 260 L of 190-proof alcohol/tonne of OD wood (Lloyd and Harris 1955).

The potential sugars from a tonne of wood are glucose (500 kg) and mannose (120 kg). The yield of glucose from the resistant cellulose is calculated as illustrated in figure F1. Of the potential 120 kg of mannose, it is assumed that 90% will end up in the product solutions. It will be primarily in the hemicellulose stream and some will be in the glucose-rich stream; but 90%, or 108 kg of mannose, is considered as available for fermentation. Of the potential glucose, 37.5 kg (7.5%) is from readily hydrolyzable cellulose, and it is supposed that 7.0% is recovered as fermentable sugar in the hemicellulose stream. These sugars, which are not from the resistant cellulose, will yield 90.8 L of ethanol. The remaining 169.2 L of ethanol must come from the resistant cellulose, which has a glucose equivalent of 462.5 kg. Consequently, conversion of the resistant cellulose to glucose must occur in 57.6% yield.

Yields from Percolation Processing of Southern Red Oak

The wood composition presumed as the basis for the percolation process is that given in table 1 (sample 3) and the same as that used in the first process calculations.. One tonne of this material contains 184 kg of xylan, equivalent to 209 kg of xylose, and 378 kg of glucan, equivalent to 420 kg of glucose (fig. F2). It is assumed that 80% of the xylose will be recovered in the hemicellulose fraction. Unlike mannose, the xylose released during the resistant cellulose hydrolysis will be largely destroyed. The hemicellulose cut will contain 7.5% of glucose originating from the readily hydrolyzable cellulose. This is not the only glucose accompanying the xylose fraction. Because of poor fractionation of the components, some of the glucose obtained from hydrolysis of the resistant cellulose will be washed out with the hemicellulose-derived sugars. It was assumed that the glucose content of the hemicellulose stream would be twice that obtained with perfect separation. The remaining glucose available for fermentation yields 124 L of ethanol.

Yields from Two-Stage Processing of Southern Red Oak

The yields indicated (fig. F3) are those previously established in the text for two-stage processing of southern red oak.

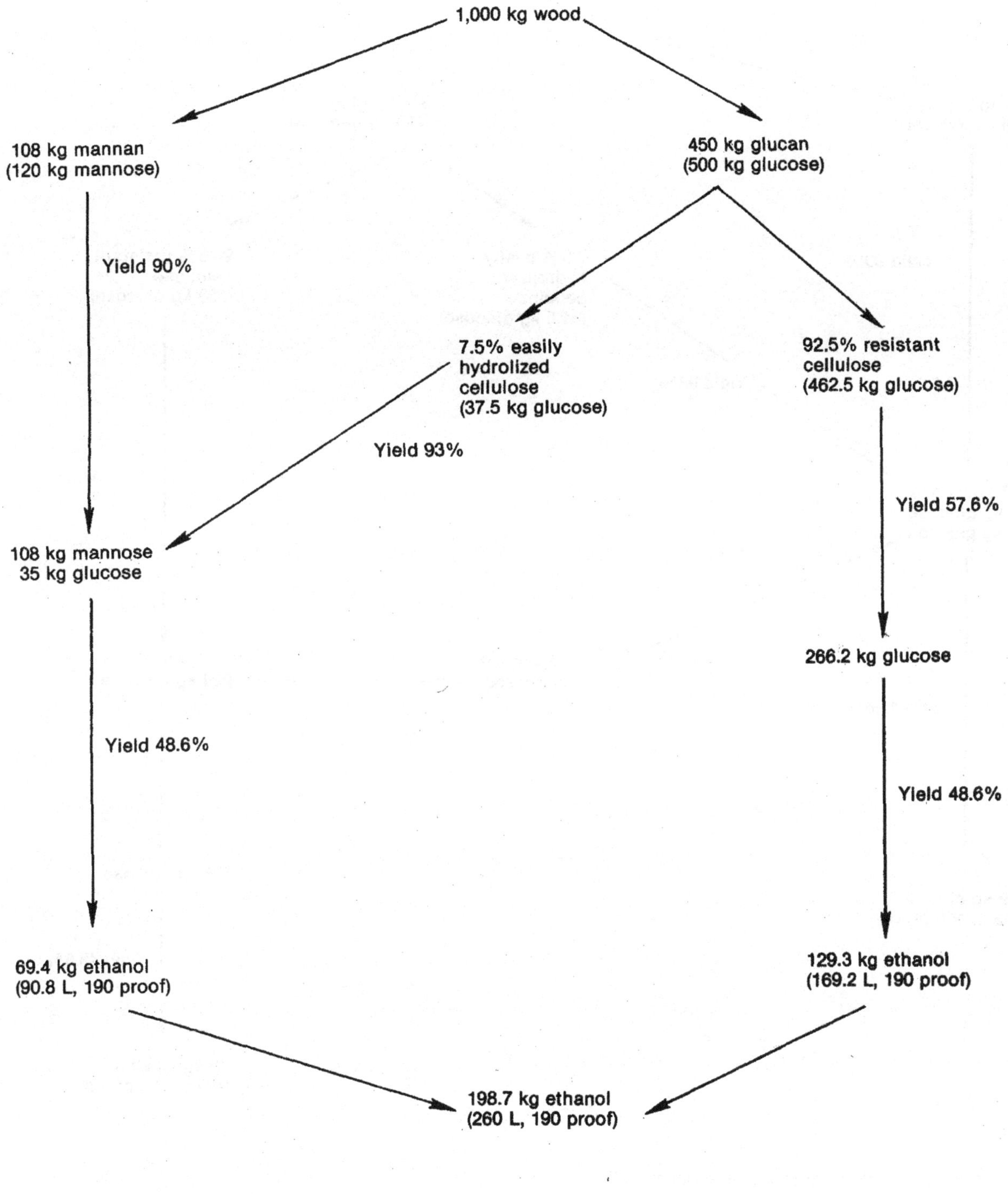

Figure F1.—Estimation of saccharification efficiency in the percolation process.

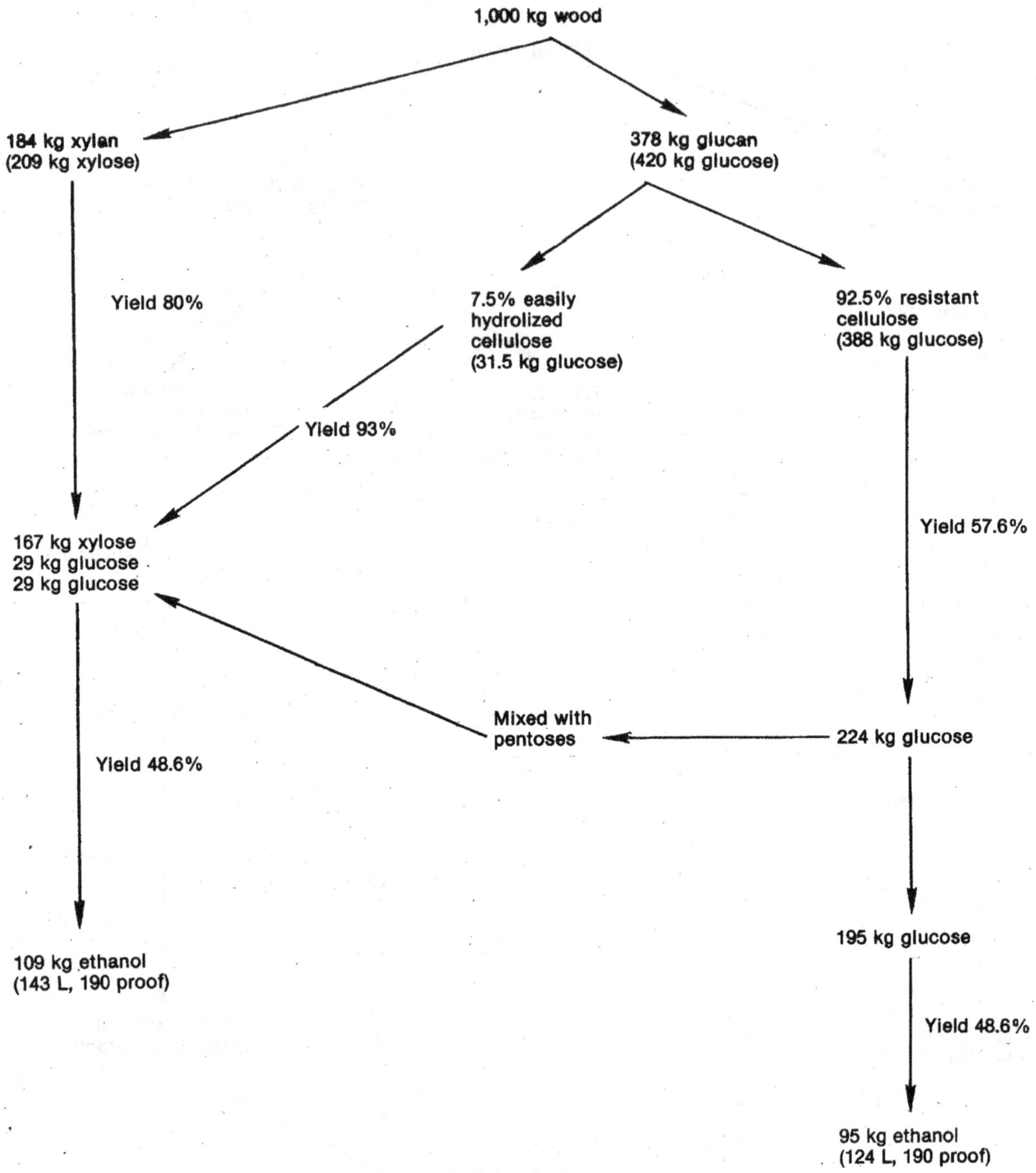

Figure F2.—Percolation processing of southern red oak.

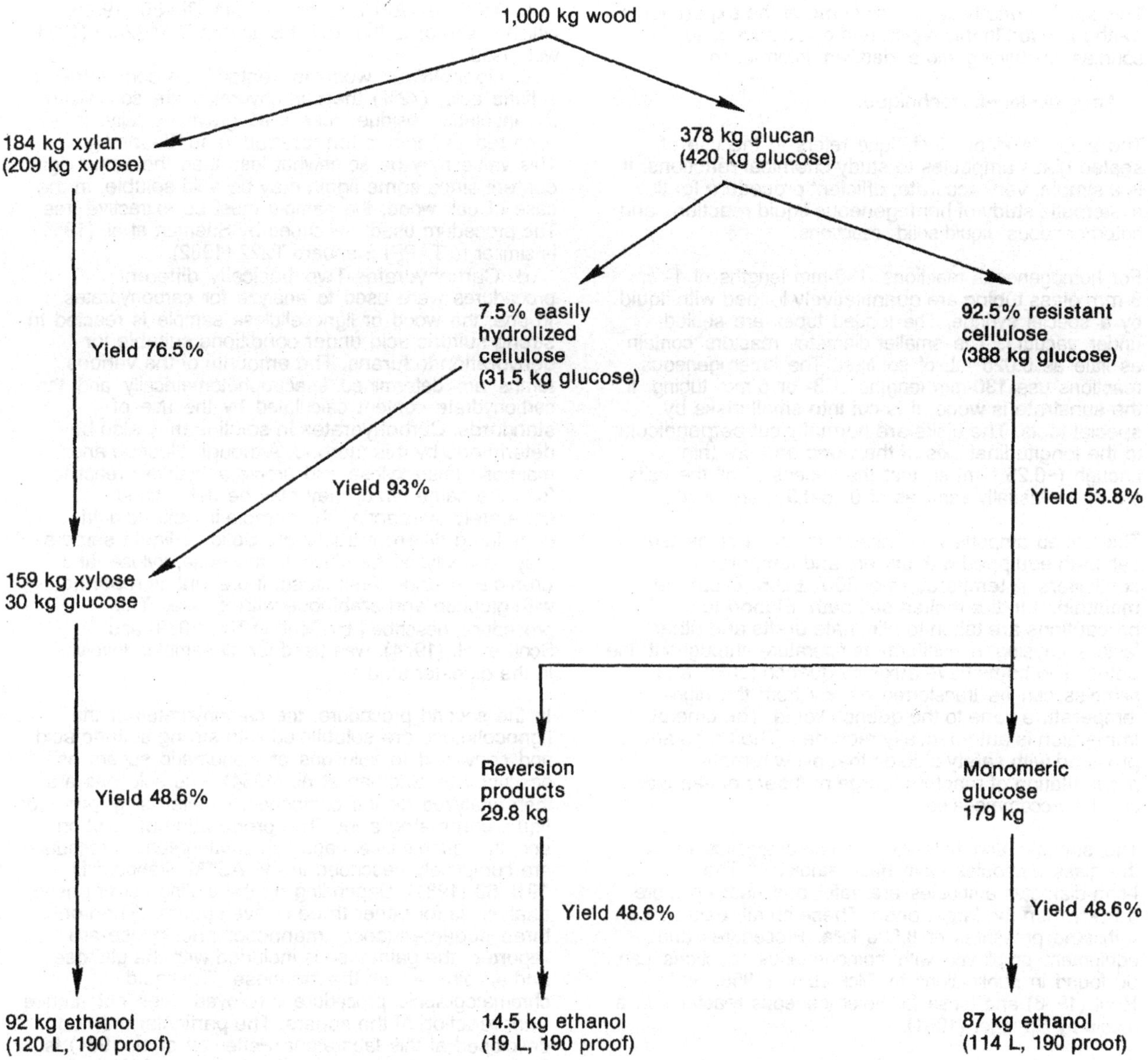

Figure F3.—Two-stage processing of southern red oak.

Appendix G
Experimental and Chemical Analysis Procedures

This section briefly describes some of the experimental methods used in this report and gives access to sources containing more detailed information.

1. Ampoule kinetic technique.

The ampoule kinetic technique refers to the use of sealed glass ampoules to study chemical reactions. It is a simple, very accurate, efficient procedure for the systematic study of homogeneous liquid reactions and heterogeneous liquid-solid reactions.

For homogeneous reactions, 130-mm lengths of 1- or 3-mm glass tubing are quantitatively loaded with liquid by a special syringe. The loaded tubes are sealed under vacuum. The smaller-diameter reactors contain as little as 0.025 mL of solution. The heterogeneous reactions use 130-mm lengths of 3- or 5-mm tubing. If the substrate is wood, it is cut into small disks by special tools. The disks are normally cut perpendicular to the longitudinal axis of the wood and are thin enough (~0.25 mm) so that the lumens of all the cells are open. Usually samples of 0.25-0.5 g are used.

The loaded ampoules are heated in an oil or molten salt bath equipped with stirrers and temperature controllers. A temperature of 300° ± 0.5° C can be maintained in the molten salt bath. Elaborate precautions are taken to eliminate drafts and other factors causing nonuniform temperature throughout the baths. The baths have auxiliary quench tanks, and samples can be transferred quickly from the high-temperature zone to the quench liquid. The time of immersion is automatically recorded. The baths are provided with safety shields that allow remote manipulation of reactors. Large numbers of samples can be accommodated.

The strength and heat-transfer characteristics of the glass ampoules have been studied.[1,2] The l-mm-diameter ampoules are safer and heat up more quickly than the larger ones. These small reactors can withstand pressures of 8,500 kPa. Procedures and equipment employed with homogeneous reactions can be found in publications by McKibbins (1958) and Root (1956) and those for heterogeneous reactions in a thesis by Springer (1961).

2. Lignocellulose and wood analyses.

a. Extractives-Wood containing tannins, such as southern red oak, require extraction with ethanol-benzene followed by extraction with ethanol (TAPPI Standard T12 (1982)). This analysis was done only on wood samples.

b. Ash-The TAPPI Standard T211 OM-80 (1982), which is similar to the ASTM Standard D 1102-56 (1981), was used.

c. Lignin-When wood Is treated with concentrated sulfuric acid, (72%) the carbohydrates are solubilized; the insoluble residue, measured gravimetrically, is reported as lignin, often referred to as Klason lignin. This value may be somewhat less than the actual lignin content since some lignin may be acid soluble. In the case of oak wood, the sample must be extractive free. The procedure used, described by Saeman et al. (1954), is similar to TAPPI Standard T222 (1982).

d. Carbohydrates-Two basically different procedures were used to analyze for carbohydrates. In one, the wood or lignocellulose sample Is reacted In strong sulfuric acid under conditions suitable for dehydration to furans. The amounts of the various furans are determined spectrophotometrically and the carbohydrate content calculated by the use of standards. Carbohydrates In solution may also be determined by this method. Although glucose and mannose (also xylose and uronic anhydride) react to form the same furan, they may be determined separately by reacting the sample in sulfuric acid containing different additives. Solid or liquid samples may be analyzed for glucose, mannose, xylose, and uronic anhydride. Galactose, if present, is reported with glucose and arabinose with xylose.. This procedure, described by Scott (1976; 1979) and Scott et al. (1974), was used for all samples Involved in the digester studies.

In the second procedure, the carbohydrates of the lignocellulose are solubilized with strong sulfuric acid and converted to solutions of monomeric sugars as described by Saeman et al. (1954). The solutions are then analyzed for the component sugars using, paper or liquid chromatography. The preparation of solutions and the quantitative paper chromatographic procedures are completely described in the ASTM Standard D 1915-63 (1981). Depending on the eluting solvent used, analysis is for either three or five sugars. When only three sugars-glucose, mannose, and xylose-are reported, the galactose is included with the glucose and arabinose with the mannose. The liquid chromatographic procedure employed does not require prior reaction of the sugars. The particular method developed at this laboratory (Pettersen et al. 1984) Is based on previous work by Wentz et al. (1982) and Paice et al. (1982).

In addition to the dehydration procedure mentioned above, wood samples were sometimes analyzed for uronic anhydride by a more conventional procedure (Browning 1949). This method relies on the quantitative liberation of CO, induced by refluxing in 12% hydrochloric acid.

---

[1]Tomlin, R.; Baumgartner, J. Strength and heat transfer tests of glass tubing reactors. Madison, WI: U.S. Forest Service, Forest Products Laboratory; May 1956. 7 p. Unpublished report.

[2]Wasser, R. B. Heat transfer tests of glass tubing reactors. Madison, WI: U.S. Forest Service, Forest Products Laboratory; Dec. 1957. 6 p. Unpublished report.

e. Acetyl–Two procedures were used for acetyl determination. One was similar to a method reported by Wiesenberger (1947). The acetyl groups are released by hydrolysis in alkaline medium; the solution is then acidified and acetic acid removed by steam stripping. The amount of isolated acid is determined by titration.

In the second procedure, the acetyl groups are hydrolyzed in the same manner as above, but the acetic acid released is assayed by gas chromatography. For this, a Supelco 60/80 Carbopak C/0.3% carbowax 20 M/0.1% H_3PO_4 column (Supelco, Inc., Bellefonte, Pa.) was used at the conditions recommended by the manufacturer. The quantity of acetic acid was determined with an internal standard of propionic acid.

3. Solution components analyses.

a. Carbohydrates-In addition to the applicable procedures discussed in the preceding section for solids, two others should be noted:

A Beckman Glucose Oxidase Analyzer (Beckman Instruments, Inc., Fullerton, Calif.) was used without modification of the recommended procedure, to determine monomeric glucose.

The Schaffer-Somogyi volumetric procedure (1933) or Nelson's calorimetric modification of Somogyi's method (1944) was used to measure the reducing power of solutions. These methods were used for solutions containing a single sugar but were also valuable for determining the amounts of oligomers in solution. This was done by measuring the increase in reducing power resulting from mild acidic hydrolysis of the sample.

b. Other components–Components were considered in two groups:

Acetic, formic, and levulinic acids-These acids were determined by high-performance liquid chromatography using a Bio-Rad HPX-87H column (Bio-Rad Laboratories, Richmond, Calif.) following the supplier's recommended operating conditions. This was the only analysis used for formic and levulinic acids, but most acetic acid samples were analyzed by the method used for acetyl.

Furfural and hydroxymethylfurfural–These compounds can be separated from other components of hydrolysates using the same HPLC procedure as used for the above acids. The few HMF analyses reported were obtained in this way. However, furfural measurements were obtained by the distillation procedure developed by Root (1956).

The Forest Products Laboratory (USDA Forest Service) has served as the national center for wood utilization research since 1910. The Laboratory, on the University of Wisconsin-Madison campus, has achieved worldwide recognition for its contribution to the knowledge and better use of wood.

Early research at the Laboratory helped establish U.S. industries that produce pulp and paper, lumber, structural beams; plywood, particleboard and wood furniture, and other wood products. Studies now in progress provide a basis for more effective management and use of our timber resource by answering critical questions on its basic characteristics and on its conversion for use in a variety of consumer applications.

Unanswered questions remain and new ones will arise because of changes in the timber resource and increased use of wood products. As we approach the 21st Century, scientists at the Forest Products Laboratory will continue to meet the challenge posed by these questions.

Forest Products Laboratory
75th
1910 1985
ANNIVERSARY

WOOD HYDROLYSIS FOR SUGAR PRODUCTION

Original report dated March 1955

Reprinted 1963

INFORMATION REVIEWED
AND REAFFIRMED 1966

No. 2029

FOREST PRODUCTS LABORATORY
MADISON 5, WISCONSIN

UNITED STATES DEPARTMENT OF AGRICULTURE
FOREST SERVICE
In Cooperation with the University of Wisconsin

WOOD HYDROLYSIS FOR SUGAR PRODUCTION[1]

By

ROGER A. LLOYD, Chemical Engineer
and
JOHN F. HARRIS, Chemical Engineer

Forest Products Laboratory,[2] Forest Service
U. S. Department of Agriculture

Introduction

Wood hydrolysis is the chemical reaction of the cellulosic components in wood with water. Acid is required to catalyze the reaction. The smallest units, or single links, derived from the cellulose chains by this reaction are sugar molecules. As many as five distinct kinds of sugar molecules are commonly obtained by hydrolysis of the carbohydrate material in wood. Glucose, which has six carbon atoms in its molecule, is the predominant sugar obtained from wood. Mannose, which also is a six-carbon sugar, is the second most common sugar obtainable from softwoods and obtained in lesser amounts from hardwoods, while xylose, a five-carbon sugar, occupies a comparable position of importance for the hardwood species. The other two common sugars derived from wood are galactose and arabinose. The carbohydrate materials in wood can be classified for hydrolysis purposes into two types--cellulose which is easily hydrolyzable and that which is resistant to hydrolysis. The easily hydrolyzable material can yield all five types of sugar. The fraction that is difficult to hydrolyze will yield largely glucose, with lesser amounts of mannose and xylose.

History

Braconnot in 1819 discovered that cellulose could be dissolved in concentrated acid solutions and converted to sugar. Much research has been carried out since then in attempts to develop economically sound processes for making sugar

---

[1] Report originally published March 1955.

[2] Maintained at Madison, Wis., in cooperation with the University of Wisconsin.

from wood. A particularly active period occurred around the turn of the century, when research and development work was in progress both in this country and in Europe. M. F. Ewen and G. H. Tomlinson are credited with developing a dilute sulfuric acid process which was used successfully in two commercial plants during the World War I era. One of the plants was located at Georgetown, S. C., and the other at Fullerton, La. The "American process," as this early hydrolysis method is now called, used a short reaction time in a rotary digester at a comparatively low temperature. As a result, the process depended almost entirely on the easily hydrolyzable material for sugar production. A low wood-sugar yield of about 22 percent resulted. The end product at both plants was ethyl alcohol, each plant producing over 5,000 proof gallons per day. The great demand for lumber depleted timber holdings rapidly, causing a curtailment of the mill operations that furnished wood waste to the hydrolysis plants. The shortage of raw material and low blackstrap molasses prices forced both plants to close soon after World War I.

Wood hydrolysis received considerable attention in Europe during the period between the Warld Wars I and II. From this era emerged two radically different processes that attained particular prominence in Germany during World War II. One used 40 percent hydrochloric acid at atmospheric pressure in expensive acid-resistant equipment. High acid requirements made acid recycling imperative and contributed markedly to excessive plant costs. High product quality and high sugar yield, however, were claimed to offset the high capital investment and production costs.

The other process, commonly known as the Scholler process, used dilute sulfuric acid and steam pressures up to 200 pounds per square inch to promote the hydrolysis reaction. It differed from the American process in that a stationary digester was used and many batches of acid liquor were passed through the same charge. From 16 to 20 hours were required to complete the hydrolysis. Sugar yields of 70 to 77 percent of the theoretical maximum were reported for the Scholler method of operation.

The Cliffs-Dow Chemical Company acquired the American rights to the Scholler process in 1935 and conducted a pilot-plant evaluation of it at Marquette, Mich. In 1943, at the request of the War Production Board, the Forest Products Laboratory began to reexamine dilute sulfuric acid hydrolysis, using the pilot-plant facilities of the Cliffs-Dow Chemical Company for the initial investigations. With information acquired at this pilot plant along with Scholler specifications, a full-scale demonstration wood-hydrolysis plant was designed and built at Springfield, Oreg. This Government-sponsored plant was not completed until after the end of World War II. Although the plant has been in partial experimental operation, sufficient capital has not been available to modify it in order to bring it into full-scale production.

Rapid-Percolation Wood Hydrolysis Process

A rapid percolation process developed by the Forest Products Laboratory is similar to the Scholler method of hydrolysis in that a dilute sulfuric acid solution is percolated through wood chips or sawdust in a stationary digester. It differs, however, in that, after an initial low-temperature hydrolysis period, the acid solution is pumped in continuously at the top and hydrolyzate is removed at the bottom with no interruption until the hydrolysis is completed. A brief description of the rapid percolation process as used with the 60-cubic-foot hydrolyzer of this Laboratory follows.

Wood waste in the form of sawdust, shavings, or chips is loaded into the hydrolyzer. When filled, the hydrolyzer is temporarily capped and steam is injected rapidly above the chip bed. This packs the bed and permits more wood to be added. The procedure is usually repeated and, after a third filling, the cap is bolted down securely.

Steam is admitted to the hydrolyzer again and allowed to pass through the bed and out the bottom to remove air and heat the chips. After steam begins to flow rapidly from the bottom vent, the valve on this line is closed and steaming is continued until the charge reaches a temperature of 293° F. At this point, the steaming operation is discontinued and the water and acid pumps are started. A calculated quantity of hydrolysis liquor containing 2.5 percent of acid is first injected to compensate for the diluting effect of the condensate from the steaming operation and moisture in the wood. The composite acid concentration inthe vessel is thus brought to 0.7 percent. The acid pump then is readjusted to provide a 0.7 percent solution. The acid solution enters the hydrolyzer at a constant rate calculated to give a 2.5:1 liquor-to-wood ratio in the hydrolyzer in 45 minutes. The temperature of this entering stream is held at 293° F. for 30 minutes and then is taken to 338° F. for the next 15 minutes.

After 45 minutes of elapsed time, the hydrolyzate control valve is opened at the bottom of the vessel and the takeoff rate is adjusted to make the liquor level coincide with the chip-bed surface level in order to control the density of the bed. The water pump is adjusted to give a hot acid-liquor flow rate equal to 0.08 times the weight of the dry charge per minute, and the acid pump is adjusted to provide a 0.6 percent acid solution. The temperature of the acid liquor is increased at the rate of 1-3/4° F. per minute, so that after a total elapsed time of 1 hour it is at the maximum of 365° F. By this time, most of the low-temperature hydrolyzate has left the vessel and hydrolysis of the resistant cellulose is under way in the upper part of the chip bed.

The high through-put rate is continued until the sugar content in the hydrolyzate stream reaches approximately 3 percent. At this time, the flow rates are cut in half, and maintained thus until the sugar concentration drops below 2 percent, when pumping of the acid is discontinued. Hydrolyzate removal continues until the liquor level in the hydrolyzer reaches a point 10 to 12 inches below the surface of the chip bed. The blow-down valve at the bottom of the vessel is then opened and the lignin residue, the solid material left in the hydrolyzer, is discharged into the lignin receiver.

These conditions are particularly suitable for Douglas-fir wood and produce, after 2-1/2 hours of operating time, a yield of sugar in the hydrolyzate equal to 65 to 67 percent of the potential sugar in the wood. About 43 pounds of sugar are obtained from 100 pounds of dry Douglas-fir wood, which has a potential sugar yield of 65 percent. The sugar concentration of the composite hydrolyzate obtained from Douglas-fir wood is roughly 5.5 percent. Douglas-fir sawdust high in bark has a potential sugar yield of 60 percent or less. This does not seem to affect the sugar yield on the basis of the potential sugar yield, but it does lower the sugar concentration of the composite hydrolyzate to a value below 5 percent.

Optimum operating conditions vary with wood species, particle size distribution, and overall charge composition; and suitable temperature schedules, pumping rates, and holding times for various wood waste materials must be determined empirically.

A Proposed Wood-Sugar Molasses Plant

A flow diagram of a proposed wood-sugar molasses plant (fig. 1) illustrates the hydrolysis equipment as well as the evaporation and neutralization equipment needed to convert the hydrolyzate into a molasses product. This plant could process 60 tons of wood (dry basis) per day in 3 hydrolyzers, each with a capacity of 5 tons of wood. The starting time of each hydrolysis cycle would be staggered in 1-1/2- to 2-hour intervals so that the weak hydrolyzate from the last portion of one hydrolysis cycle could be flashed directly into a second hydrolyzer and be used as the initial hydrolysis liquor for the cycle just starting. Recycling of a portion of the hydrolyzate would affect heat, acid, and lime economy. Four cycles per hydrolyzer, or a total of 12 cycles, are possible each 24 hours in this equipment.

Table 1 lists the equipment needed for such a plant and approximate 1954 costs. The acid-resistant materials of construction markedly affect the cost of much of the equipment. From these costs the following capital investment can be calculated:

Total Estimated Investment Cost[3]

(A) Total installed equipment cost		$423,600
Process piping	20% of (A)	84,700
Instrumentation	5% of (A)	21,200
Building	10% of (A)	42,100
(B) Total physical plant cost		571,600
Engineering and construction overhead	20% of (B)	114,200
Contingencies	20% of (B)	114,200
	Total estimated investment cost	800,000

Table 2 shows the estimated production costs, exclusive of wood costs. Wood-sugar molasses costs in cents per gallon as a function of the price of the wood waste at the plant, assuming a conservative yield of sugar in the molasses product equal to 35 percent of the dry wood waste, are as follows:

Wood cost delivered-- dollars per ton (dry basis)	0	4	8
Production cost-- cents per gallon	20.6	23.4	26.2

As can be seen in figure 1, the proposed plant uses lime for neutralizing the acid in the hydrolyzate. If ammonia was used for neutralizing instead of lime, neutralization equipment could be simplified, evaporator maintenance reduced, and protein value added to the molasses product. Neutralization with ammonia would, however, increase the production cost values given above by approximately 1/2 cent per gallon.

The importance of plant size can be illustrated by recalculating the production costs for plants of other sizes. Production costs of molasses, as given above, would be 3 to 4 cents a gallon less for a 120-ton-per-day plant and 6 to 8 cents per gallon more for a 30-ton plant.

---

[3]Steam and water facilities would add approximately $100,000 to $120,000 if built specifically for this plant.

Economic Picture for Wood-Sugar Molasses

From October 1952 to October 1954, the price of blackstrap molasses, a by-product of cane sugar manufacture, was comparatively stable, varying from a low of 10.1 to a high of 12.7 cents per gallon F.O.B. tank car New York. At New Orleans, the price was 1 to 2 cents lower, and at Minneapolis the price to brokers and feed mixers was approximately 9 cents higher than the New York prices. It is apparent, from these data and the production costs of table 2, that wood-sugar molasses, in order to compete with blackstrap, must be produced for minimum cost by employing a large-scale operation and low-cost raw material, and in an area where transportation costs are low for the wood-sugar product and high for blackstrap. Such locations, where a large, continuing supply of low-cost wood waste is available and where a good market is assured, are few in this country.

Evaluation of Wood-Sugar Molasses as an Animal Feed

Over a 3-year period beginning in 1948, an extensive program for evaluating wood-sugar molasses as an animal feed was carried out at nine State Agricultural Experiment Stations. Over 200 tons of molasses were prepared in the pilot plants of the Tennessee Valley Authority and the U. S. Forest Products Laboratory for this testing program. Although most of the tests were concerned with dairy or beef cattle, sheep, pigs, and poultry were also used at some stations. From the results of these many experiments, the conclusion is that, in general, wood-sugar molasses is equivalent to blackstrap molasses as a carbohydrate feed supplement. Summary data of the individual tests are available in another Forest Products Laboratory publication.[4]

Wood-sugar molasses and blackstrap differ in many respects even though they are fairly comparable nutritionally. Blackstrap molasses thickens at low temperatures to a greater extent than the product from wood. This is a factor affecting the pumping and other flow properties of the material.

In a softwood molasses such as Douglas-fir, approximately 15 percent of the sugars are pentoses and 85 percent are hexoses fermentable to alcohol. More than 30 percent of the sugars in hardwood molasses can be nonfermentable pentoses. More than 90 percent of the sugars in blackstrap can be fermentable and about half are present as sucrose.

The total solids content in blackstrap molasses is approximately 80 percent, as compared with 60 to 65 percent for wood-sugar molasses. A large part of this difference is due to the high ash content in blackstrap--roughly 10 percent, as compared to 2 to 3 percent for molasses from wood.

---

[4]Harris, Elwin E. Hydrolysis of Wood for Stock Feed. Forest Products Laboratory Report R1731. 24 pp. Illus. (Processed).

The continual precipitation of solids from wood-sugar molasses is a fault that is not common to blackstrap molasses. The sediment, which seems to form in greater quantity in hardwood molasses, tends to plug drum openings, pipe lines, and nozzles, and its physical nature is such that it entraps a significant quantity of sugar. Some control of this problem can be effected by inexpensive clarification methods at the molasses plant. Sediment formation is a progressive phenomenon that becomes more troublesome the longer the molasses is in storage; therefore, if wood-sugar molasses is used within a month after preparation, any tendency for solids to precipitate should be negligible.

Complete removal of the precipitate-forming substances from the sugar, however, would involve ion-exchange processing, which could add as much as 8 cents a gallon to production costs in a 60-ton-per-day plant.

Other Uses for Wood Sugar

Wood sugar can be converted into a host of chemical products. Some of the various processes will be put to use whenever and wherever the economics is sound. Two pulp mills in Canada and one in this country have for several years been successfully fermenting and making ethyl alcohol from the wood sugars present in their spent sulfite pulping liquors. This is possible because a low-cost value can be assigned to these sugars, since they were formerly discarded and created a pollution problem.

The cost of producing wood sugars by an acid hydrolysis process would be lower if the lignin residue from the process had chemical value and could thus offset a part of the production costs. Lignin is a potential source of reactive compounds that, if efficiently produced, could make this byproduct worth more than as a fuel. When a practical method for producing reactive lignin or lignin products is integrated with the hydrolysis of the cellulosic material in wood, the resulting process should have an important role in wood-waste utilization.

Ethyl alcohol, which up to a few years ago, was produced entirely by fermentation processes, is more and more produced by synthesis from petroleum byproducts. Today only one major company still makes alcohol by fermentation. To make fermentation competitive, a continuing supply of a low-cost fermentable sugar is needed. Except during a national emergency, sugar from the hydrolysis process cannot be considered for this use unless its cost is reduced by very efficient use of the lignin and the nonfermentable sugars from the hydrolysis process.

Butyric, acetic, and lactic acids, butyl and isopropyl alcohols, and 2, 3-butylene glycol and glycerine are fermentation products that have been produced experimentally from wood-sugar solutions, but none are now being produced from this material on a commercial scale.

The development within recent years of a very efficient continuous yeast propagation process has made yeast production one of the more promising possibilities for wood sugar. Commercial plants in both this country and abroad are so engaged. The plants in this country are operating on the wood sugars present in the spent liquors from sulfite pulping. The product, which has high protein and vitamin content, is used largely for animal feed preparations.

The yeast process converts efficiently both the 5- and 6-carbon sugars to an insoluble protein material that can be concentrated inexpensively in centrifuges and with a minimum of processing can be made ready for marketing. This process would be suitable for converting the entire sugar yield from a wood hydrolysis process; or it would function well as a secondary process to utilize the unused sugars from a main operation, such as an ethyl alcohol fermentation.

Crystalline sugar can be produced from the hydrolyzate of the rapid percolation process; however, the nonsugar solids content of this material is high, and the yield of sugar crystals is low unless the crystallization is preceded by ion-exchange purification. It would be imperative that a crystallization process for wood sugar be integrated with other processes that would use the byproducts efficiently.

Furfural and levulinic acid, which are derived, respectively, from pentose and hexose sugars by acid hydrolysis, hold much promise for becoming important chemicals from wood, and research is currently in progress at this Laboratory on processes for their production.

Lignin

The lignin residue is an important byproduct of the wood-hydrolysis process, since it is roughly equal to one-third the weight of the dry wood charge. The most likely use for this material to date is as a fuel to produce steam for the process. After it is flashed in the lignin receiver, the residue contains approximately 60 percent moisture and 8,300 British thermal units per pound on a moisture-free basis, and can furnish over 60 percent of the heat required for the entire wood-sugar molasses process.

Because of its apparent low reactivity, the lignin residue from the rapid-percolation wood hydrolysis process has found no important chemical application to date; however, by the use of chemical techniques such as hydrogenation or oxidation, it is quite possible that, in the not-too-distant future, lignin will be the raw material for many important chemical compounds.

Table 1.--Estimated cost of a wood hydrolysis plant with capacity of 60 tons per day[1]

Item	Quantity	Description	Cost
1	2	4 ton/hr. hogs, complete with pneumatic conveyor	
2	1	Bulldozer for log handling	$ 60,000
3	1	Cyclone separator, capacity 500 cu. ft.	
4	3	Chip receivers with strain gauge weighing device, capacity 800 cu. ft.	
5	3	Hydrolyzer vessel - 7 ft. dia. by 20 ft. high, 200 lb. per sq. in. working pressure -- Carbon steel clad with Everdur	86,400
6	3	Hydrolyzer blowdown valves, 8-inch diameter -- phosphorus bronze	30,000
7	1	Lignin receiver 12 ft. dia. by 14 ft. high	3,600
8	3	Flash tanks 6 ft. dia. by 6 ft. high	3,500
9	1	Hydrolyzate blending tank 12 ft. dia. by 14 ft. high, complete with mixer.	3,200
10	1	Sulfuric acid storage tank 8 ft. dia. by 20 ft. high	1,900
11	1	Water heater	1,400
12	3	Flash condensers -- 135 sq. ft. each	2,700
13	1	Water supply pump	3,700
14	1	Water heater circulating pump	900
15	2	Concentrated acid supply pumps	2,200
16	1	Hydrolyzate settling tank 12 ft. dia. by 8 ft. high	2,200
17	3	Molasses settling tanks 12 ft. dia. by 12 ft. high complete with mixers	4,000
18	2	Slurry and neutralizing tanks 5 ft. dia. by 5 ft. high, complete with mixers	1,500
19	2	Rotary filters -- 50 sq. ft. each	15,000
20	3	Calandria type evaporators, complete with vacuum equipment and pumps	201,000
21	1	Circulating filter pump	400
		Total installed plant cost	423,600

[1]All costs -- 1954 level (Engineering News-Record Index-550)

Table 2.--Production costs for plant with capacity of 60 tons of wood waste a day

Cost item	Total cost				Costs of wood preparation		Costs of hydrolysis		Costs of evaporation	
	Annual requirement	Unit price	Cost per year	Percent of total	Cost per year	Percent of total	Cost per year	Percent of total	Cost per year	Percent of total
Direct production costs										
Raw materials										
Wood tons	18,000	......	......	......	......	......	......	......	......	......
H_2SO_4-66° Be tons	960	$25.00	$ 24,000	4.62	......	......	$ 24,000	4.62	......	......
Slaked lime tons	645	15.00	9,700	1.87	......	......	......	......	$ 9,700	1.87
Utilities										
Steam M-lbs.	243,000	.67	156,000	30.00	......	......	41,000	7.89	115,000	22.11
Water M-gals.	450,000	.031	14,000	2.69	......	......	2,000	.39	12,000	2.30
Electricity M-kw. hrs.	1,200	6.4	7,700	1.48	$ 5,000	0.96	1,700	.33	1,000	.19
Operating labor hrs.	36,000	2.00	72,000	13.85	18,000	3.46	27,000	5.19	27,000	5.20
Supervision	......	......	7,200	1.38	1,800	.35	2,700	.52	2,700	.51
Payroll overhead	......	......	15,800	3.04	4,000	.77	5,900	1.13	5,900	1.14
Maintenance	......	......	40,000	7.69	5,700	1.10	12,900	2.48	21,400	4.11
Supplies	......	......	6,000	1.15	900	.17	1,900	.37	3,200	.61
Total direct production costs	......	......	352,400	67.78	35,400	6.81	119,100	22.91	197,900	38.06
Indirect production costs										
Technical and analytical services	......	......	71,500	13.75	15,300	2.94	25,600	4.92	30,600	5.89
Fixed costs										
Depreciation	......	......	80,000	15.39	11,300	2.17	25,800	4.96	42,900	8.26
Local taxes and insurance	......	......	16,000	3.08	2,300	.44	5,200	1.00	8,500	1.64
Total production costs	......	......	519,900	100.00	64,300	12.37	175,700	33.79	279,900	53.84

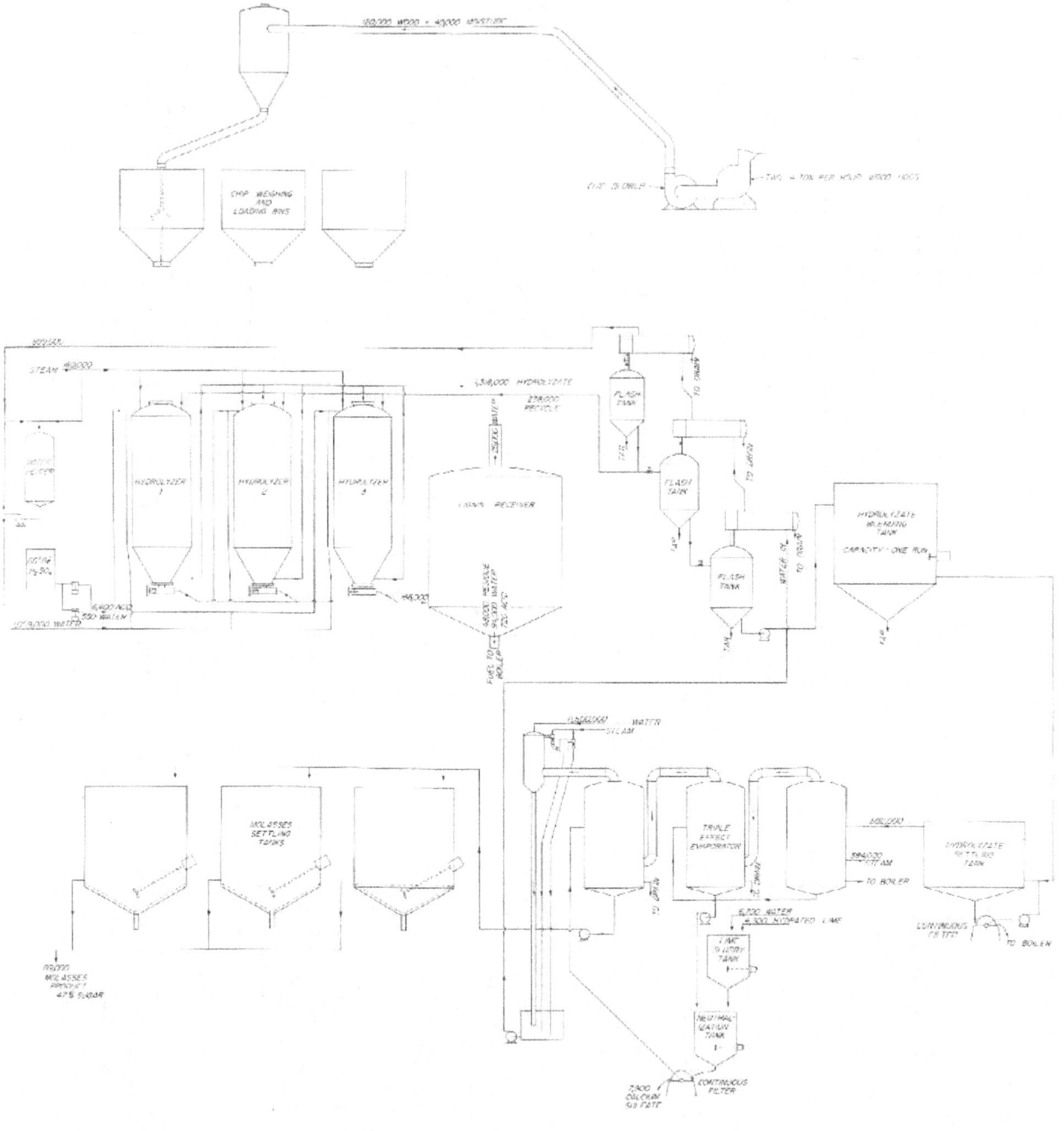

Figure 1.--A wood-sugar molasses plant processing 60 tons of dry wood per 24 hours. All quantities in pounds per 24 hours.

Z M 96097 F

SUBJECT LISTS OF PUBLICATIONS ISSUED BY THE FOREST PRODUCTS LABORATORY

The following are obtainable free on request from the Director, Forest Products Laboratory, Madison 5, Wisconsin.

- List of publications on Box and Crate Construction and Packaging Data
- List of publications on Chemistry of Wood and Derived Products
- List of publications on Fungus Defects in Forest Products and Decay in Trees
- List of publications on Glue, Glued Products, and Veneer
- List of publications on Growth, Structure, and Identification of Wood
- List of publications on Mechanical Properties and Structural Uses of Wood and Wood Products
- Partial list of publications for Architects, Builders, Engineers, and Retail Lumbermen
- List of publications on Fire Protection
- List of publications on Logging, Milling, and Utilization of Timber Products
- List of publications on Pulp and Paper
- List of publications on Seasoning of Wood
- List of publications on Structural Sandwich, Plastic Laminates, and Wood-Base Aircraft Components
- List of publications on Wood Finishing
- List of publications on Wood Preservation
- Partial list of publications for Furniture Manufacturers, Woodworkers and Teachers of Woodshop Practice

Note: Since Forest Products Laboratory publications are so varied in subject, no single list is issued. Instead a list is made up for each Laboratory division. Twice a year, December 31 and June 30, a list is made up showing new reports for the previous 6 months. This is the only item sent regularly to the Laboratory's mailing list. Anyone who has asked for and received the proper subject lists and who has had his name placed on the mailing list can keep up to date on Forest Products Laboratory publications. Each subject list carries descriptions of all other subject lists.

FPL FILING SYSTEM DESIGNATION--WC 5

Lloyd, Roger Albert
Wood hydrolysis for sugar production, by R. A. Lloyd and J. F. Harris. Madison, Wis., U.S. Forest Products Laboratory, 1955 (Reprinted 1963)
8 p., illus. (F.P.L. rpt. no. 2029)

A process for producing wood-sugar molasses is discussed, including a description of equipment needed, total investment cost, and uses for wood sugar.

Lloyd, Roger Albert
Wood hydrolysis for sugar production, by R. A. Lloyd and J. F. Harris. Madison, Wis., U.S. Forest Products Laboratory, 1955 (Reprinted 1963)
8 p., illus. (F.P.L. rpt. no. 2029)

A process for producing wood-sugar molasses is discussed, including a description of equipment needed, total investment cost, and uses for wood sugar.

Lloyd, Roger Albert
Wood hydrolysis for sugar production, by R. A. Lloyd and J. F. Harris. Madison, Wis., U.S. Forest Products Laboratory, 1955 (Reprinted 1963)
8 p., illus. (F.P.L. rpt. no. 2029)

A process for producing wood-sugar molasses is discussed, including a description of equipment needed, total investment cost, and uses for wood sugar.

Lloyd, Roger Albert
Wood hydrolysis for sugar production, by R. A. Lloyd and J. F. Harris. Madison, Wis., U.S. Forest Products Laboratory, 1955 (Reprinted 1963)
8 p., illus. (F.P.L. rpt. no. 2029)

A process for producing wood-sugar molasses is discussed, including a description of equipment needed, total investment cost, and uses for wood sugar.

Made in the USA
[illegible]
06 November 2016

Made in the USA
Monee, IL
08 November 2026

67931557R00057